Michell Mario Reimer

Motor neuron regeneration in the spinal cord of adult zebrafish

Michell Mario Reimer

Motor neuron regeneration in the spinal cord of adult zebrafish

(HAMILTON, 1822)

Südwestdeutscher Verlag für Hochschulschriften

Impressum/Imprint (nur für Deutschland/ only for Germany)
Bibliografische Information der Deutschen Nationalbibliothek: Die Deutsche Nationalbibliothek verzeichnet diese Publikation in der Deutschen Nationalbibliografie; detaillierte bibliografische Daten sind im Internet über http://dnb.d-nb.de abrufbar.

Alle in diesem Buch genannten Marken und Produktnamen unterliegen warenzeichen-, marken- oder patentrechtlichem Schutz bzw. sind Warenzeichen oder eingetragene Warenzeichen der jeweiligen Inhaber. Die Wiedergabe von Marken, Produktnamen, Gebrauchsnamen, Handelsnamen, Warenbezeichnungen u.s.w. in diesem Werk berechtigt auch ohne besondere Kennzeichnung nicht zu der Annahme, dass solche Namen im Sinne der Warenzeichen- und Markenschutzgesetzgebung als frei zu betrachten wären und daher von jedermann benutzt werden dürften.

Verlag: Südwestdeutscher Verlag für Hochschulschriften Aktiengesellschaft & Co. KG
Dudweiler Landstr. 99, 66123 Saarbrücken, Deutschland
Telefon +49 681 37 20 271-1, Telefax +49 681 37 20 271-0
Email: info@svh-verlag.de
Zugl.: Hamburg, Universität, Diss., 2008

Herstellung in Deutschland:
Schaltungsdienst Lange o.H.G., Berlin
Books on Demand GmbH, Norderstedt
Reha GmbH, Saarbrücken
Amazon Distribution GmbH, Leipzig
ISBN: 978-3-8381-1706-5

Imprint (only for USA, GB)
Bibliographic information published by the Deutsche Nationalbibliothek: The Deutsche Nationalbibliothek lists this publication in the Deutsche Nationalbibliografie; detailed bibliographic data are available in the Internet at http://dnb.d-nb.de.

Any brand names and product names mentioned in this book are subject to trademark, brand or patent protection and are trademarks or registered trademarks of their respective holders. The use of brand names, product names, common names, trade names, product descriptions etc. even without a particular marking in this works is in no way to be construed to mean that such names may be regarded as unrestricted in respect of trademark and brand protection legislation and could thus be used by anyone.

Publisher: Südwestdeutscher Verlag für Hochschulschriften Aktiengesellschaft & Co. KG
Dudweiler Landstr. 99, 66123 Saarbrücken, Germany
Phone +49 681 37 20 271-1, Fax +49 681 37 20 271-0
Email: info@svh-verlag.de

Printed in the U.S.A.
Printed in the U.K. by (see last page)
ISBN: 978-3-8381-1706-5

Copyright © 2010 by the author and Südwestdeutscher Verlag für Hochschulschriften Aktiengesellschaft & Co. KG and licensors
All rights reserved. Saarbrücken 2010

1	**INTRODUCTION**	1
1.1	Aims of the study	1
1.2	Zebrafish (*Danio rerio*) as a model organism	1
1.3	Mammals, including humans, do not regenerate the lesioned or diseased CNS.	2
1.4	Anamniotes (amphibians and fish) have a high regenerative capacity, which includes the CNS	3
1.5	The zebrafish shows anatomical and functional spinal cord regeneration	3
1.6	Primary motor neurons in developing zebrafish provide a model for studying motor axon differentiation	5
1.7	Cell recognition molecules in axonal pathfinding	6
1.8	Summary	7
2	**MATERIAL AND METHODS**	8
2.1	Materials	8
2.1.1	Enzymes	8
2.1.2	Bacterial strains	8
2.1.3	Bacterial media	9
2.1.4	Vectors	9
2.1.5	Kits	9
2.1.6	DNA Standards	10
2.1.7	Oligonucleotides	10
2.1.8	Primers	10
2.1.9	Morpholinos	11
2.1.10	Antibodies	11
2.1.11	Buffers and solutions	12
2.1.12	Chemicals	13
2.1.13	Equipment	13
2.1.14	Zebrafish	14
2.2	Molecular biological methods	15
2.2.1	Standard Polymerase chain reaction (PCR)	15
2.2.2	Nested PCR	16

2.2.3	Touchdown PCR	16
2.2.4	TA cloning	16
2.2.5	TOPO cloning	17
2.2.6	Purification of DNA fragments and PCR products	17
2.2.7	Restricition enzyme digestion of DNA	17
2.2.8	Agarose gel electrophoresis of DNA fragments	17
2.2.9	Dephosphorylation of DNA fragments	18
2.2.10	Ligation of DNA fragments	18
2.2.11	Transformation of plasmid DNA into bacteria	18
2.2.12	Miniprep (small scale plasmid preparation)	18
2.2.13	Midiprep (medium scale plasmid preparation)	19
2.2.14	Quantification of DNA	19
2.2.15	Sequencing of DNA	19
2.2.16	Precipitation of DNA	19
2.2.17	Total RNA extraction from zebrafish tissue	20
2.2.18	First strand cDNA synthesis	20
2.2.19	*In vitro* transcription	20
2.3	Histological Methods	21
2.3.1	Immunohistochemistry on cryosections	21
2.3.2	*In situ* hybridisation on cryosections	22
2.3.3	Immunohistochemistry on vibratome sections of adult spinal cord	23
2.3.4	Stereological quantifications in spinal cord sections	24
2.3.5	Profile counts in spinal cord sections	24
2.3.6	Microinjection into zebrafish eggs	25
2.3.7	Whole mount immunohistochemistry	25
2.3.8	Whole mount *in situ* hybridisation	26
2.4	Animal experiments	27
2.4.1	Perfusion fixation of adult zebrafish	27
2.4.2	Spinal cord lesion of adult zebrafish	27

2.4.3	Retrograde tracing of adult zebrafish	27
2.4.4	Intraperitoneal substance application	28
3	**RESULTS**	**29**
3.1	Adult spinal cord regeneration	29
3.1.1	Spinal cytoarchitecture is not restored in a spinal lesion site	29
3.1.2	A spinal lesion triggers ventricular proliferation	30
3.1.3	Motor neurons show significant regenerative capacity	33
3.1.4	*Olig2:GFP⁺* ependymo-radial glial cells are potential motor neuron stem cells in the adult spinal cord	42
3.1.5	Expression of ventral neural tube markers is increased in a developmentally appropriate pattern	48
3.1.6	Cyclopamine inhibits *shh* dependent motor neuron regeneration	52
3.2	Motor axon pathfinding during development	55
3.2.1	Cloning of *plexinA3*	55
3.2.2	*PlexinA3* is strongly expressed in spinal motor neurons	57
3.2.3	*PlexinA3* is necessary for motor axon pathfinding	58
3.2.4	*PlexinA3* morpholino phenotypes are specific	61
4	**DISCUSSION**	**63**
4.1	Adult zebrafish are capable of motor neuron regeneration	63
4.2	*Olig2⁺* ependymo-radial glial cells are the putative stem cells in adult motor neuron regeneration	66
4.3	Mechanisms of motor neuron regeneration in adult zebrafish are similar to developmental mechanisms	67
4.4	Implications of motor neuron regeneration in zebrafish for spinal cord regeneration in mammals	68
4.5	*PlexinA3* is crucial for motor axon pathfinding	69
4.6	Conclusion	70
5	**SUMMARY**	**71**
6	**LITERATURE**	**73**
7	**APPENDIX**	**81**
7.1	Abbreviations	81

7.2　Morpholinos 82
7.3　Overexpression-construct *plexinA3* 83
7.3.1　Primers used to clone *plexinA3* overexpression construct: 83
7.3.2　Sequence of the overexpression construct for *plexinA3*:....................83
7.3.3　Restriction enzyme map for *plexinA3* overexpression construct 87
7.4　Publications...88

1 INTRODUCTION

1.1 Aims of the study

(aim 1) Adult zebrafish, in contrast to mammals, show an amazing capacity for functional spinal cord repair (Kirsche, 1950; Becker et al., 1997; van Raamsdonk et al., 1998; Becker et al., 2004). However, cellular regeneration of spinal neurons, such as motor neurons has not been analysed. Therefore, this study asks whether motor neurons that are lost due to spinal injury regenerate in adult zebrafish and if so, what are the cellular and molecular mechanisms of neuronal regeneration.

(aim 2) In order to analyse the molecular mechanism of axonal differentiation of motor neurons, which may be recapitulated during regeneration, the well established system of axon growth from so-called primary motor neurons in embryonic zebrafish was used (Beattie, 2000). It has been shown that cell recognition molecules are important for axon growth and pathfinding (Beattie, 2000; Giger et al., 2000; Feldner et al., 2005). Therefore, this study asks which specific cell recognition molecules are necessary for correct growth of primary motor axons during embryonic development.

Together, these aims are intended to increase our understanding of motor neuron differentiation in general and during successful regeneration of the adult spinal cord in particular. Ultimately, insights from zebrafish into these evolutionarily conserved mechanisms may help to cure human conditions, such as spinal cord injury and motor neuron disease.

1.2 Zebrafish (*Danio rerio*) as a model organism

The zebrafish (*Danio rerio*) is part of the family of *Cyprinidae*. It belongs to the class of *Actinopterygii*, in the infraclass of the *Teleostei*. These 2-4 cm long freshwater fish can be found in South Asia, Northern India, Bhutan, Pakistan and Nepal.

The genome of the zebrafish is partially duplicated in evolution (Taylor et al., 2001). Therefore a substantial number (up to 30%) of mammalian genes have two orthologs in the zebrafish genome. Conveniently, the zebrafish genome is now fully sequenced (www.ensembl.org, Sanger Institute), providing easy access to gene sequences in-silico. This facilitates the design of transgenic reporter lines, which are relatively easy to generate in zebrafish, and of morpholinos (antisense-oligo nucleotides) for gene knock-down studies. The possibility to inject RNA overexpression constructs as well as morpholinos from the one cell stage egg and the transparancy of embryos makes the zebrafish an ideal model system for studying developmental processes in vivo (Nasevicius and Ekker, 2000; Malicki et al., 2002).

Furthermore, zebrafish development is well characterised and a variety of transgenic reporter lines that express fluorescent proteins in motor neurons as well as antibodies that label motor neurons are available (Renoncourt et al., 1998; Higashijima et al., 2000; William et al., 2003; Flanagan-Steet et al., 2005).

1.3 Mammals, including humans, do not regenerate the lesioned or diseased CNS.

CNS injury or disease in mammals often causes irreversible loss of motor and sensory function (Dijkers, 2005). The properties of axonal regeneration and its failure in mammals has been extensively studied. It is thought that the lack of axonal regeneration in mammals is due to inhibitory molecules, such as myelin-associated inhibitors (e.g. nogo-A), myelin-associated glycoprotein, and oligodendrocyte myelin glycoprotein, that prevent axon outgrowth (Spencer et al., 2003; Schwab, 2004). Other inhibitory molecules are part of the extracellular matrix, such as chondroitin sulfate proteoglycans, which are found in the glial scar (Carulli et al., 2005). Another reason for regeneration failure is inflammation, which often leads to a further increase of damage to the CNS (Bambakidis et al., 2008).

The regeneration and replacement of lost neurons in adult mammals is not so well-characterised. Studies have show that neuronal progenitor cells in the subventricular zone and dentate gyrus in the adult mammalian brain proliferate and differentiate into neurons (Johansson, 2007). In the spinal cord of rats, proliferation and differentiation of glial progenitor cells that give rise to astrocytes and oligodendrocytes has been demonstrated (Horner et al., 2000) but neurogenesis has never been observed. To find the signalling pathways that trigger endogenous progenitor cells to differentiate into neurons after a lesion or disease and replace lost neurons could be one way to ameliorate the devastating effects after CNS damage.

1.4 Anamniotes (amphibians and fish) have a high regenerative capacity, which includes the CNS

The zebrafish is well established as a model in developmental studies and interest in adult regeneration, e.g. of heart tissue (Poss et al., 2002) and spinal cord (Becker et al., 1997), is increasing.

Zebrafish show an impressive tissue regeneration capacity at the adult stage (Bernhardt, 1999; Becker et al., 2004). After a spinal cord lesion, zebrafish grow an axonal bridge between the two ends of the fully transected spinal cord and regain swimming function (Becker et al., 2004). In tail regeneration paradigms in amphibians in which the tail, including the spinal cord is amputated (Echeverri and Tanaka, 2002; Beck et al., 2003) the tail is regenerating from an advancing blastema. This includes a completely regenerated spinal cord.

1.5 The zebrafish shows anatomical and functional spinal cord regeneration

In 1950, Walter Kirsche described in detail the morphological response to a complete spinal cord transection in adult teleosts (*Poecilia reticulata*) (Kirsche, 1950). Based on his morphological observations that large "ganglion cells" disappeared and later reappeared, he even hypothesized the replacement of lost neurons in response to a lesion event.

A complete transection of the spinal cord leads to loss of movement in the distal body part. Swimming performance in zebrafish has been tested in a tunnel with a constant water flow. Swimming behaviour recovered after a lesion and plateaued around 2.5 months post-lesion (van Raamsdonk et al., 1993). However, while significant recovery of swimming behaviour occurred, performance of the fish remained worse than in unlesioned fish. Another method to quantify functional recovery after a lesion is to measure spontaneous movement in an open-field setup. This test shows a recovery in swim distance at 6 weeks post-lesion, which was indistinguishable from sham (muscle)-lesioned controls (Becker et al., 2004). The difference in the results of these test paradigms may be that to be forced to swim against a flow is more challenging for the fish than to perform their normal swimming patterns. This indicates that while regeneration after spinal injury occurs, it is not perfect.

Anatomically, long-range axonal projections are destroyed after a complete transection. These include descending axons from the brainstem, intraspinal descending connections, ascending axons from dorsal root ganglia and intraspinal neurons providing sensory feedback to the brainstem. Substantial regrowth of spinal axons is only observed from brainstem neurons. Blocking the regrowth of these long-range axonal projections abolishes the capacity for functional recovery of the adult zebrafish (Becker et al., 2004). While this clearly demonstrates that axonal regrowth from the brainstem is essential for functional recovery after spinal cord injury, the plastic changes in the spinal cord, e.g. regeneration of target neurons, remain largely unknown.

The signalling cascade leading to motor neuron differentiation during development is well understood and is evolutionarily conserved. Progress in recent years in identifying extracellular signals and cell-intrinsic differentiation programs has led to a general model of early generation of different classes of neurons. Most of this data were obtained from studies with chick and mouse embryos, but motor neuron differentiation is very similar in embryonic zebrafish (Park et al., 2002). Generally, a gradient of the morphogen *sonic hedgehog* (*Shh*) regulates the expression of a set of transcription factors in progenitor cells of the ventral spinal cord. The pattern of transcription factor expression defines five domains of progenitor cells, termed p0, p1, p2, pMN and p3. Specific cell types are produced from each domain, leading to the generation of different types of interneurons and somatic motor neurons. The pMN domain gives rise to motor neurons (Fig. 1). Specifically, a high concentration of *sonic hedgehog* in combination with the transcription factors *nkx6.1*, *pax6* and *olig2* define the motor neuronal cell fate in progenitor cells (Jessell, 2000; Briscoe and Ericson, 2001; Shirasaki and Pfaff, 2002). The resulting immature neurons are positive for the motor neuron marker *islet1/2* and *HB9* (Higashijima et al., 2000; Flanagan-Steet et al., 2005).

The question arises, whether some or all of these mechanisms are re-capitulated during adult motor neuron regeneration.

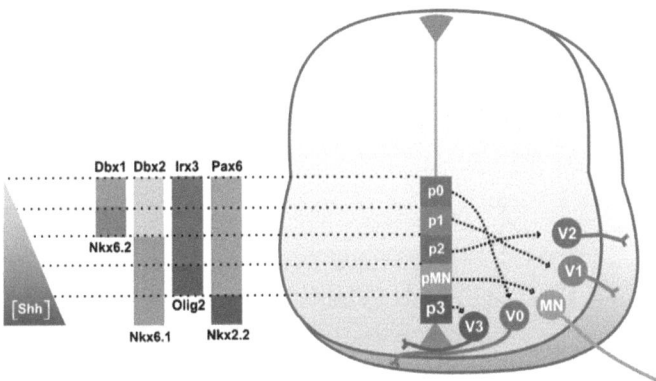

Fig. 1: Spinal cord neurons. Schematic diagram of the development of early classes of ventral spinal cord neurons in mice. A gradient of the morphogen *Sonic hedgehog* (*Shh*) regulates the expression of a set of transcription factors in progenitor cells of the ventral spinal cord. The pattern of transcription factor expression defines five domains of progenitor cells, termed p0, p1, p2, pMN, and p3. A specific cell type is produced from each domain, leading to the generation of V0, V1, V2, and V3 interneurons and somatic motor neurons (MN). After (Kullander, 2007).

1.6 Primary motor neurons in developing zebrafish provide a model for studying motor axon differentiation

A widely used model system to study signals for early motor axon growth, is the outgrowth of primary motor axons in zebrafish embryos (Beattie, 2000). This is because there are only three primary motor neurons per spinal hemi-segment. These neurons grow axons out of the spinal cord following a common path in the middle of each segment to the horizontal myoseptum. From there the axons paths diverge. The axon of the caudal primary motor neuron (CaP) grows towards the ventral somite, pioneering the ventral motor nerve. The axon of the middle primary motor neuron (MiP) follows the CaP axon up to the horizontal myoseptum where it retracts and grows towards the dorsal somite. The rostral primary motor neuron (RoP) axon takes a lateral direction from the horizontal myoseptum (Fig. 2). In some of the hemisegments a variable primary motor neuron (VaP) is present, which sometimes develop beside the CaP and mostly die from interaction with the CaP (Eisen et al., 1986; Myers et al., 1986; Westerfield et al., 1986; Eisen et al., 1990; Sato-Maeda et al., 2008).

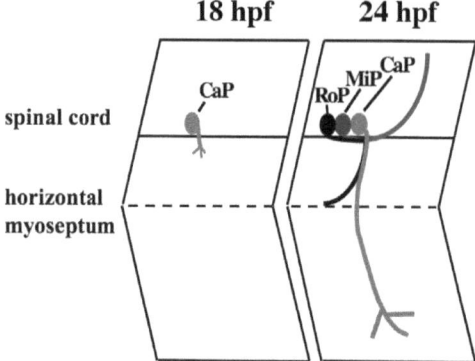

Fig. 2: Schematic illustration of primary motor axon outgrowth in embryonic zebrafish. A side view of zebrafish trunk segments at 18 and 24 hpf is given. At 18 hpf the caudal primary motor neuron (CaP) grows an axon out of the spinal cord. At 24 hpf, the axons of the middle (MiP) and rostral (RoP) primary motor neurons have followed on the common pathway to the horizontal myoseptum and the MiP has grown towards the dorsal somite. The CaP axon is the only one growing ventrally beyond the horizontal myoseptum.

1.7 Cell recognition molecules in axonal pathfinding

The molecular mechanisms underlying axonal pathfinding are pathway cues and axonal receptors. PlexinA1 to A4 are co-receptors for axon-repelling or attracting class 3 extracellular semaphorins. It is thought that neuropilin-1 (NRP1) or neuropilin-2 (NRP2) is the ligand-binding part and plexins are the signal transducing part of semaphorin class 3 receptors (for recent review, see Kruger et al., 2005). Removing individual components from this guidance network leads to specific defects of nerve growth (Giger et al., 2000; Huber et al., 2005; Yaron et al., 2005), indicating distinct roles for different ligand/receptor combinations in the pathfinding of different axon populations.

Sema3Aa and sema3Ab (zebrafish homologs of mammalian sema3A) are expressed in the trunk environment. Overexpression of either ligand reduces growth of primary motor axons (Roos et al., 1999; Halloran et al., 2000) and anti-sense morpholino oligonucleotide knockdown of sema3A1 leads mainly to aberrant branching of the CaP axon (Sato-Maeda et al., 2006). Knockdown of neuropilin-1a (NRP1a) alone or in double knockdown experiments with semaphorin ligands leads to nerve branching, additional exit points of axons from the spinal cord, and ventral displacement of neuronal somata along the extra-spinal motor axon pathway (Feldner et al., 2005). This suggests that semaphorins guide primary motor axons by repellent mechanisms via NRP1a containing axonal receptors. So far, the role of plexins has not been examined. The only class A member of the plexin family characterised in zebrafish is plexinA4, but it is not expressed in primary trunk motor neurons (Miyashita et al., 2004). Therefore, we have investigated the role of another plexinA, *plexinA3* in the outgrowth of primary motor axons. The knowledge of embryonic neurogenesis and axonal outgrowth of motor neurons may lead to further insights into mechanisms of adult spinal cord regeneration.

1.8 Summary

In this study I demonstrate for the first time that adult zebrafish are capable of regenerating motor neurons lost after spinal cord lesion. Evidence is provided that these neurons fully differentiate and are integrated into the spinal network. I identify the morphogen *shh* as one of the signals that is important for motor neuron differentiation and progenitor cell proliferation at the adult stage. Embryonic studies indicate that the cell recognition molecule *plexinA3* is pivotal for correct motor axon pathfinding. These findings provide insight into the differentiation processes of motor neurons, both in development and regeneration in a vertebrate.

2 MATERIAL AND METHODS

2.1 Materials

2.1.1 Enzymes

Restriction endonucleases various (5-20 U/µl)	New England Biolabs UK Ltd. (Hitchin, Hertfordshire, UK)
DNA polymerase	
PfuUltra™ HF DNA Polymerase	Stratagene (Amsterdam, NL)
Taq DNA Polymerase with Standard Taq Buffer	New England Biolabs UK Ltd. (Hitchin, Hertfordshire, UK)
Reverse Transcriptases	
SuperScript II™ RT	Invitrogen (Karlsruhe, D)
SuperScript III™ RT	Invitrogen Ltd. (Paisley, UK)
RNaseOUT™ Recombinant Ribonuclease Inhibitor	Invitrogen Ltd. (Paisley, UK)
RNasin®Plus RNase Inhibitor	Promega (Mannheim, D)
Miscellaneous	
T4 DNA Ligase	New England Biolabs UK Ltd. (Hitchin, Hertfordshire, UK)
Alkaline Phosphatase, shrimp (SAP)	Roche (Mannheim, D)
Alkaline Phosphatase	Roche Diagnostics Ltd. (Burgess Hill, UK)
Proteinase K	Roche Diagnostics Ltd. (Burgess Hill, UK)

2.1.2 Bacterial strains

XL1-Blue competent cells	Stratagene (UK)

NEB Turbo Competent *E. coli* (High Efficiency)	New England Biolabs UK Ltd. (Hitchin, Hertfordshire, UK)
dam⁻/dcm⁻ Competent E. coli	New England Biolabs UK Ltd. (Hitchin, Hertfordshire, UK)
E. Coli One Shot®TOP10	Invitrogen (Karlsruhe, D)
E. Coli DH5α	Invitrogen (Karlsruhe, D)

2.1.3 Bacterial media

All bacterial media were autoclaved before use. If nessecary Ampicillin or Kanamycin was added.

Bacterial growth media encapsulated media LB medium	QBIOgene, Fisher Scientific (UK)
LB Agar Miller Fisher BioReagents	Fisher Scientific (UK)
Antibiotics	
Ampicillin (50mg/ml in H_2O stock, 50µg/ml working solution)	
Kanamycin (50mg/ml in H_2O stock, 30µg/ml working solution)	

2.1.4 Vectors

pGEM®-T easy	Promega (Southamton, UK)	TA cloning vector
pCR®-Blunt II-TOPO®	Invitrogen (UK)	TOPO cloning vector
pCS2+MT	(Rupp et al., 1994)	mRNA overexpression vector
pBlueScript® II	Stratagene (UK)	

2.1.5 Kits

MEGAscript™ (T3/T7/SP6)	Ambion (Cambridge, UK)
mMESSAGE mMACHINE™	Ambion (Cambridge, UK)
Poly (A) Tailing Kit	Ambion (Cambridge, UK)
Rapid DNA Ligation Kit	Roche Diagnostics Ltd. (Burgess Hill,

	UK)
pGEM®-T easy vector system I	Promega (Southamton, UK)
Zero Blunt® TOPO® PCR Cloning Kit	Invitrogen (UK)
QIAquick™ Gel Extraction	Qiagen (Crawley, UK)
QIAquick™ PCR Purification	Qiagen (Crawley, UK)
MiniElute™ Gel Extraction Kit	Qiagen (Crawley, UK)
MiniElute™ PCR Purification Kit	Qiagen (Crawley, UK)
HiSpeed® Plasmid Midi Kit	Qiagen (Crawley, UK)
RNeasy® Midi Kit	Qiagen (Crawley, UK)
GFX™ Micro Plasmid Prep Kit	GE Healthcare (Little Chalfont, UK)
High Pure PCR Product Purification Kit	Roche (Mannheim, D)

2.1.6 DNA Standards

GeneRuler™DNA Ladder Mix, ready to use	Fermentas (York, UK)
Ready-Load™ 1Kb Plus DNA Ladder	Invitrogen (UK)
Quick-Load® 2-Log DNA Ladder(0.1-10.0 kb)	New England Biolabs UK Ltd. (Hitchin, Hertfordshire, UK)
100 bp DNA Ladder	New England Biolabs UK Ltd. (Hitchin, Hertfordshire, UK)
DNA sample buffer (10x)	Eppendorf (UK)

2.1.7 Oligonucleotides

Primer (purification: desalted)	metabion (Martinsried, D)
Primer (purification: RP-Column)	TAGN Ltd (Gateshead, UK)
Primer (purification: RP-Column)	VH Bio Ltd . (Gateshead, UK)

2.1.8 Primers

plexin A3 (BamHI) forward
5`- GTGGATCCATGAGGTCCTTGTGGCTG -3`

plexinA3 (BamHI) reverse

5`- TAGGATCCGCTGCTGCCAGACATCAG-3`
olig2 forward
5`- TCCAGCAGACCTTCTTCTCC -3`
olig2 reverse
5`- ACAACTGGACGGATGGAAACC -3`
patched 1 forward
5`- GTCTGCAAGCCACTTTTGATGC -3`
patched 1 reverse
5`- GGGGTAGCCATTGGGATAGT -3`
GAPDH forward
5`- ACTCCACTCATGGCCGTT -3`
GAPDH reverse
5`- TCTTCTGTGTGGCGGTGTAG -3`

2.1.9 Morpholinos
Synthetic antisense oligonucleotides (morpholino) were used to "knockdown" genes, Blocking eighter the translation of the mRNA or the splicing of the preRNA. Morpholinos were synthesised by Gene Tools LLC (Philomath, OR, USA), sequences see appendix.

2.1.10 Antibodies

anti-*HB9* /*MNR2* (81.5C10)	Dr. T.M. Jessell (Columbia University, New York, USA), 1:400, Developmental Studies Hybridoma Bank (Tanabe et al., 1998)
anti-*islet-1/-2* (40.2D6)	Developmental Studies Hybridoma Bank(Iowa City, USA), 1:1000 (Tsuchida et al., 1994)
anti-acetylated tubulin (6-11B-1)	Sigma Aldrich (UK), 1:1000
anti-NCAM-PSA (735)	Prof. Dr. Rita Gerardy-Schahn (MHH, Hannover) 1:1000 (Kibbelaar et al., 1989)
anti-neurofilament-associated antigen (3A10)	Dr. T.M. Jessell (Columbia University, New York, USA), 1:50, Developmental Studies Hybridoma Bank

anti-myc epitope (9E10)	Santa Cruz Biotechnology, (Santa Cruz, USA), 1:600
rat anti-BrdU (BU 1/75)	AbD Serotec (Oxford, UK) 1:500
anti-*PCNA* (PC10)	Dako Cytomation (Glostrup, Denmark) 1:500
anti-*nkx6.1* (AB2024)	O.Madsen (Hagedorn Research Institute, Gentofte, Denmark)1:1000
anti-GFP (A11122)	Invitrogen (UK) 1:200
pax6 (MiniPerm 95)	Veronica van Heyningen (MRC, Edinburgh)

All Cy2-, Cy3-, Cy5 and HRP conjugated anti-rabbit, anti-rat and anti-mouse secondary antibodies were from Jackson ImmunoResearch Laboratories Inc. (West Grove, PA, USA) or Dianova (Hamburg, Germany),1:200. Goat Serum (ab7481) was used for blocking in immunohistochemistry, (Abcam,Cambridge, UK) and heat inactivated prior use for 30min at 60°C.

2.1.11 Buffers and solutions

Method-specific solutions that are not listed below are specified in the corresponding chapters.

blocking buffer (whole mount immunohistochemistry)	1x PBS 1% (v/v) DMSO 1% (v/v) normal goat serum (NGS) 1% (w/v) BSA 0.7% (v/v) Triton-X 100
blocking buffer (vibratome section immunohistochemistry)	1.5% (v/v) normal goat serum in PBSTx
blocking solution (whole mount in-situ hybridisation)	1% (w/v) blocking reagent (Boehringer) in PBST
Citrate buffer	10mM sodium citrate in 1x PBS, pH 6.0
DAB stock solution	6% (w/v) diaminobenzidine (DAB)

Danieau solution	58 mM NaCl
	0.7 mM KCl
	0.4 mM MgSO$_4$
	0.6 mM Ca(NO$_3$)$_2$
	5 mM HEPES
	pH 7.6
dNTP stock solution (100mM)	dATP, dCTP, dGTP, dTTP, 25 mM each
phosphate buffer saline (10x PBS)	1.36 M NaCl
	0.1 M Na$_2$HPO$_4$ 27 mM KCl
	18 mM KH$_2$PO$_4$
	pH 7.4
PBST	0.1% (v/v) Tween 20 in 1x PBS
PBStx	0.1% (v/v) Triton X 100 in 1x PBS
PFA	4% paraformaldehyde (w/v) in 1 xPBS
Saline sodium citrate buffer (SSC) (20x stock)	3 M NaCl
	0.3 M tri-sodium citrate
	pH 7.4
Tris-acetate-EDTA buffer (TAE) (50x stock)	2M Tris-acetate
	100mM EDTA
	pH 8.5

2.1.12 Chemicals

Chemicals were purchased as *pro analysis* quality from Sigma-Aldrich (UK) and Fisher Scientific (UK).

2.1.13 Equipment

Apotome	Zeiss (Goettingen, D)
Axiophot	Zeiss (Goettingen, D)
Bench-top centrifuges 5417 R and 5804 R	Eppendorf (Hamburg, D)
Centrifuge RC 5C Plus Sorvall	Kendro (Hanau, D)
Centrifuge Sigma 3K30C	Sigma Laborzentrifugen GmbH (Osterode am Harz, D)

Cryostat CM3050	Leica (Bensheim, D)
E.A.S.Y. UV-light documentation	Herolab (Wiesloh, D)
Fishsystem	Aqua Schwarz (Goettingen, D)
Hotplate stirrer Fisherbrand® metal top	Fisher Scientific (UK)
Hybridizer UVP HB-1000	Jencons PLS (East Grinstead, UK)
Incubated shaker MaxQ Mini 4450	Fisher Scientific (UK)
Laser scanning microscope LSM510	Zeiss (Goettingen, D)
Microcentrifuge 5415 D	Eppendorf (Hamburg, D)
Microinjector Narishige Intracel + manipulator	Intracel Ltd. (Herts, UK)
MJ mini gradient thermal cycler	Biorad (UK)
MJ PTC-200 DNA ENGINE™ Peltier Thermal Cycler	Biozym (Hessisch Oldendorf, D)
Qualicool incubator 260	LTE Scientific Ltd (Oldham, UK)
Spectrophotometer Ultrospec 3000/DPV	APB (Freiburgh, D)
Sub-Cell GT / Power Pac Basic System	Biorad (UK)
Technico Mini centrifuge	Fisher Scientific (UK)
Vibratome Microm	Optech Scientific Instruments (Oxfordshire, UK)
Wide Mini-Sub Cell GT / Power Pac Basic System	Biorad (UK)

2.1.14 Zebrafish

Zebrafish (*Danio rerio*) were kept at 26.5°C, 14-hour light and 10-hour dark cycle. They were fed two times a day, with dry flakes, ZM pellets (ZM Ltd., UK) and *Artemia salina* larvae. The fish were breed and raised according to standard protocols (Westerfield, 1989; Nusslein-Volhard).

2.2 Molecular biological methods

Standard molecular biological methods were carried out according to (J Sambrook et al., 1989) unless otherwise indicated.

2.2.1 Standard Polymerase chain reaction (PCR)

The standard PCR (Saiki et al., 1985), an amplification of DNA by *in vitro* enzymatic replication, was performed in an MJ mini-gradient thermal cycler (Biorad, UK).

Reagents:

Template (cDNA, gDNA, Plasmid DNA)	10pg – 1ng
dNTPs	200 µM (each dNTP)
Primer (forward)	0.1 – 1 µM
Primer (reverse)	0.1 – 1 µM
Reaction buffer (10x)	1x
DNA Polymerase (1min/kb Taq DNA Polymerase, 2min/kb PfU Ultra DNA Polymerase)	2.5U
add ddH$_2$O to final volume 50 µl	

Program:

cycles	time	temperature
1	5 min	94 °C
	30 s	94 °C
25 - 40	45 s	Tm – 1 °C
	1 min per kb	72 °C
1	10 min	72 °C

Usually the reaction was carried out in a 0.2 ml PCR reaction tube. Taq polymerase was routinely used for the amplification of up to 2 kb long DNA fragments. Proof reading PfuUltra™ HF DNA Polymerase was used to amplify DNA for overexpression and full-length constructs. After the PCR reaction was finished, 5 µl of the product was analysed by agarose gel electrophoresis.

2.2.2 Nested PCR

The nested PCR approach was used to amplify sequences from genomic DNA (gDNA). A very low number of copies of a specific DNA template, e.g. a regulatory sequence from gDNA, leads often to the amplification of the wrong DNA sequence. This approach prevents the amplification of the wrong product by sequentially using two primer pairs for the same sequence. The first primer pair includes the sequence of the second primer pair and the first PCR reaction is used as a template of the second (1:40 dilution). The reaction mix is equal to the standard PCR.

2.2.3 Touchdown PCR

Another modification of the standard PCR to reduce non-specific amplicons is touchdown PCR: starting the PCR program using a higher annealing temperature than the optimum in early PCR cycles. At every cycle the annealing temperature was decreased by 1 °C until Tm − 1 °C was reached. At that temperature 20 additional cycles were performed to allow the enrichment of the wanted product over any non-specific product.

Program:

cycles	time	temperature
1	5 min	94 °C
	30 s	94 °C
20	30 s	Tm + 14 (first) °C
	30 s	Tm − 1 (last) °C
	1 min per kb	72 °C
	30 s	94 °C
20	30 s	Tm − 1 °C
	1 min per kb	72 °C
1	10 min	72 °C

2.2.4 TA cloning

DNA, obtained using *Taq* DNA polymerase, contains a single 3'-adenosine overhang to each site of the PCR product. These PCR products can directly be cloned into a linearized vector with a 3'-thymidine overhang. For this ligation reaction T4 DNA ligase is used (pGEM®-T easy vector system I, Promega).

2.2.5 TOPO cloning

PCR with PfuUltra™ HF DNA Polymerase leads to a product without any overhang. Such PCR fragments with a blunt-end were ligated in the pCR-BluntII TOPO vector, using the Zero Blunt® TOPO® PCR Cloning Kit (Invitrogen, UK).

2.2.6 Purification of DNA fragments and PCR products

Silica-matrix based columns to purify DNA (MiniElute™ PCR Purification Kit, QIAquick™ PCR Purification and High Pure PCR Product Purification Kit) were used according to manufacturer's protocol. The DNA was eluted in 50µl ddH$_2$O.

2.2.7 Restricition enzyme digestion of DNA

Double stranded DNA was digested with appropriate amounts of restriction enzymes (NEB) according to the manufacturer's protocols. Control digestions were carried out in a 20 µl total volume reaction for 2-3 hours. For preparative digestions the total reaction volume was scaled up to 100 µl overnight at the recommended temperature.

2.2.8 Agarose gel electrophoresis of DNA fragments

To separate and analyse restriction digestions and PCR products, horizontal agarose gel electrophoresis was perfomed. Gels (0.8-1.5% w/v) were prepared by heating agarose (Fisher Scientific, UK) in 1x TAE buffer. The concentration was chosen depending on the size of the DNA sample to be separated. Before pouring the gel, Ethidium-bromide (Fisher Scientific, UK) was added (7µl/100ml). For loading the samples, loading buffer (Eppendorf, UK) was mixed to a final concentration of 1x fold. Electrophoresis was performed with 10 V/cm in BIORAD gel chambers with 1x TAE buffer. For documentation, pictures were taken in an E.A.S.Y. UV-light documentation system and if necessary, bands were cut out with a scalpel.

To extract DNA from agarose gels, the QIAquick™ Gel Extraction or MiniElute™ Gel Extraction Kit from Quiagen, UK was used according to the manufacturer´s protocol.

2.2.9 Dephosphorylation of DNA fragments

To prevent linearized DNA from religating in a ligation reaction using T4 DNA ligase, the 5'-phosphates of the DNA were removed. 1U of alkaline shrimp phosphatase (Roche) dephosphorylates approximately 50 ng of linearized DNA in 20 minutes at 37 °C.

2.2.10 Ligation of DNA fragments

To ligate DNA fragments into a vector, e.g. subcloning, 50 ng vector DNA was mixed with 5x molar amount of insert DNA for blunt end or 3x molar amount for sticky end ligation. 1 µl T4 DNA ligase and 2 µl of 2x reaction buffer was added to a final reaction volume of 20 µl. Incubation was performed for 2 hours at room temperature or overnight at 16 °C.
Alternatively, the Rapid DNA ligation kit (Roche Diagnostics Ltd. Burgess Hill, UK) was used according to manufacturer´s protocol.
After the ligation, an aliquot was directly used for transfomation in *E.coli*.

2.2.11 Transformation of plasmid DNA into bacteria

2-10 µl of the ligation mix or 0.5 µl of a plasmid DNA preparation was used to transform heat shock competent *E.coli*. The DNA was added to 100 µl of the competent bacteria in a 1.5ml reaction tube and gently mixed, incubated for 30 minutes on ice, followed by a 45-second heat shock at 42 °C. After the heat shock 800 µl LB medium were added and the tube was incubated on ice for 2 minutes. Further incubation for 1 hour on a shaker at 200 rpm at 37 °C was followed by plating the bacterial solution on LB agar plates with the required antibiotic. Bacterial colonies were picked after 12-14 hours at 37 °C.

2.2.12 Miniprep (small scale plasmid preparation)

One picked colony was transfered into a 15 ml reaction tube containing 5 ml of LB medium with the required antibiotic. After incubation overnight at 200 rpm and 37 °C the plasmid was cleaned-up with the GFX™ Micro Plasmid Prep Kit (GE Healthcare).

2.2.13 Midiprep (medium scale plasmid preparation)
For large scale plasmid preparation, one colony was picked off from the LB agar plate and transfered into a sterile 250 ml Erlenmeyer flask with 50 ml LB medium. The LB medium contained the required amount of antibiotics (ampicillin 50 – 100 µg/ml, kanamycin 30 µg/ml). After overnight incubation in a shaker with 200 rpm at 37 °C, the plasmid was harvested using HiSpeed® Plasmid Midi Kit (Qiagen, Crawley, UK), according to the manufacturer´s protocol.

2.2.14 Quantification of DNA
The quantification of DNA samples was carried out directly in the aqueous solution by measuring the adsorption at a wavelenth of 260 nm against blank (aquatous solution without DNA). An optical density (OD) of 1 absorption equals approximately 50g/ml dsDNA. Alternatively, the concentration was defined using agarose gel electrophoresis with a DNA mass ruler (Quick-Load® 2-Log DNA Ladder, NEB).

2.2.15 Sequencing of DNA
For sequencing, DNA samples were sent to the Sequencing Service, College of Life Sciences, MSI/WTB Complex University of Dundee, UK. Samples were prepared according the facilities protocols (http://www.dnaseq.co.uk) and obtained using their web interface.

2.2.16 Precipitation of DNA
Sodium acetate (3M, pH 4.9, 1:10 v/v) and 2.5x volumes cold (-20 °C) ethanol absolute were added to the DNA. After mixing, the reaction tubes were kept on ice for 30 minutes and centrifuged for 15 minutes at 16000x g (RT). The supernatant was removed and the pellet was washed with 800 µl ethanol 70%. After centrifugation and removal of the supernatant the pellet was washed

repeatedly with ethanol 70% in 400 µl and 200 µl. The pellet was dried for 15 minutes and resuspended in ddH$_2$O.

2.2.17 Total RNA extraction from zebrafish tissue

To extract total RNA from brain and spinal cord tissue or whole embryos, the animals were killed via a schedule 1 method (Home Office, UK). The tissue was removed quickly and total RNA was obtained using the RNeasy® Midi Kit (Qiagen, Crawley, UK), according to the manufacturer's protocol.

2.2.18 First strand cDNA synthesis

First strand synthesis was carried out using the SuperScript III™ RT and the RNaseOUT™ Recombinant Ribonuclease Inhibitor (Invitrogen Ltd.,Paisley, UK) according to the manufacturer's protocols. The reaction steps were performed in a MJ mini-gradient thermal cycler (Biorad, UK).

total RNA	11 µl
random primers	1 µl
dNTP mix (10mM)	1 µl in 0.2ml PCR reaction tubes
mix and spin down	
5 min	65 °C
1 min	on ice, spin down
5x First strand buffer	4 µl
DTT (0.1M)	1 µl
RNase OUT	1 µl
Super Script III	1 µl
pipette to mix and spin down	
5 min	25 °C
60 min	50 °C
15 min	70 °C, store at -20 °C

2.2.19 *In vitro* transcription

To generate DIG labelled probes for *in situ* hybridisation, an *in vitro* transcription was performed using the MEGAscript™ Kit (Ambion, Cambridge, UK). 10 µg of plasmid DNA containing the wanted insert, flanked by a T3, T7 or SP6 promotor were digested with restriction endonucleases overnight. Thereby, only the promotor sequence and the desired DNA insert was transcribed. The digested plasmid DNA was precipitated as described in Precipitation of DNA (see above). For the generation of DIG labelled RNAs, DIG-11-dUTP (Roche, UK) was used instead of UTP provided by Ambion. Alternatively, e.g. for double labeling experiments, fluorescein labelled RNA probes were used. For this purpose, reactions were carried out using Fluorescein-12-UTP (Roche, UK) instead of DIG-11-UTP.

DIG-UTP mix (10x)
10 mM ATP
10 mM CTP
10 mM GTP
6.5 mM UTP
3.5 mM DIG-11-dUTP (Roche, UK)

To generate mRNA for overexpression studies, Ambion's mMESSAGE mMACHINE™ Kit (Ambion, Cambridge,UK) was used. In both cases, 20 µl *in vitro* transcriptions were performed according to the manufacturer's protocol. After the incubation time, the template DNA was destroyed by adding 1 µl DNase to the reaction mix and incubating it for 15 min at 37°C. Generated RNAs were purified by lithium chloride precipitation (part of the Kit) and stored at –80°C.

2.3 Histological Methods

2.3.1 Immunohistochemistry on cryosections

Immunohistochemistry on 14 µm cryosections was performed as described (Becker and Becker, 2001). Sections were cut on a cryostat and mounted on poly-L-lysine (0.1% PLL) covered glass slides. After drying for 10 min up to a few hours the sections were encircled with Pap Pen and fixed in Methanol at -20°C for 10 min. A single wash in PBS to remove the Methanol was followed by 30 min blocking in PBS with goat serum (15 µl serum / ml) in a wet chamber.

Then the sections were incubated in the primary antibody in PBS at 4°C in a humid chamber overnight. The following day the unbound antibody was removed by washing in PBS 3 times for 15 min and detected with the secondary antibody for 45 min at RT. Finally, 3 times washing in PBS removed the unbound antibody and mounted with Elvanol (DuPont,Wilmington, Delaware, USA).

2.3.2 *In situ* hybridisation on cryosections

Non-radioactive detection of mRNAs was performed in 14 μm cryosections. The sections were cut from freshly frozen tissue on a cryostat and mounted on glass slides, dried for maximally 45 mins and fixed in 4% PFA overnight. The next day, sections were washed 3 times in 1x PBS, treated with 0.1 M HCl for 20 min, acetylated in 0.1 M triethanolamine containing 0.25% acetic anhydride, and dehydrated in an ascending ethanol series. Finally, sections were air-dried and prehybridized for 3 hours at 37°C with hybridization mix. Hybridization with the DIG-labelled probes was performed at 55°C overnight in humid chambers. DIG-labelled probes were diluted 1:500 - 1:1000 in hybridization buffer. After hybridization, the sections were washed twice in 0.2x SSC at 55°C, followed by three washing steps in 0.2x SSC containing 50% formamide (each 90 minutes at 55°C). To prevent unspecific binding, sections were incubated in blocking buffer for 30 min prior to the antibody detection. Anti-Digoxigenin-AP antibodies (Roche, Mannheim, D), diluted 1:2000 in blocking buffer, were applied and incubated overnight at 4°C. To remove unbound antibody, sections were washed twice in Buffer1 for 15 min. The Buffer1 was removed and the sections were equilibrated for 5 min with BCIP/NBT tablets (Sigma-Aldrich) and developed with the same staining solution until signals became visible under a stereomicroscope. Finally, sections were washed in 1x PBS and coverslipped with Elvanol.

Hybridisation buffer:
25 ml deionized formamide
5 ml 10x "Grundmix"
3.3 ml 5M NaCl
2.5 ml 2M DTT
10 ml dextransulfate
4.7 ml RNase free H_2O

10x "Grundmix":
2 ml 1 M Tris pH 7.5
200 µl 0.5 M EDTA
2 ml 50x Denhardt's solution
2 ml tRNA (25 mg/ml)
1 ml poly A^+ RNA (10 mg/ml)
2.8 ml RNase free H_2O

Buffer 1:
100 mM Tris
150 mM NaCl
pH 7.5

Blocking buffer:
1% (w/v) Blocking Reagent
0.5% (w/v) BSA
in Buffer 1

2.3.3 Immunohistochemistry on vibratome sections of adult spinal cord

Immunohistochemistry on vibratome sections was carried out with rat anti-BrdU (BU 1/75, 1:500, AbD Serotec, Oxford, UK), mouse anti-*islet-1/-2* (Tsuchida et al., 1994) (40.2D6, 1:1000, Developmental Studies Hybridoma Bank, Iowa City, USA), mouse anti-*HB9* (MNR2, 1:400, Developmental Studies Hybridoma Bank) mouse anti-*PCNA* (PC10, 1:500, Dako Cytomation, Glostrup, Denmark) and goat anti-*ChAT* (AB144P, 1:250, Chemicon, Temecula, USA) antibodies.

Secondary Cy3-conjugated antibodies were purchased from Jackson ImmunoResearch Laboratories Inc. (West Grove, PA, USA). Animals were transcardially perfused with 4% paraformaldehyde and post-fixed at 4°C overnight. Spinal cords were dissected and floating sections (50 µm thickness) were produced with a vibrating blade microtome (Zeiss, Goettingen, D). Antigen retrieval was carried out by incubating the sections for 1 hour in 2 M HCl for BrdU immunohistochemistry, or by incubation in citrate buffer (10mM sodium citrate in PBS, pH=6.0) at 85°C for 30 minutes for *HB9*, *islet-1/-2* and *PCNA* immunohistochemistry. All other steps were carried out in PBS (pH 7.4) containing 0.1% triton-X100. Sections were blocked in goat serum (15 µl/ml) for 30 minutes, incubated with the primary antibody at 4° overnight, washed three times 15 minutes, incubated with the appropriate secondary antibody for 1h, washed again, mounted in 70% glycerol and analysed using a confocal microscope (Zeiss Axioskop LSM 510). Double-labeling of cells was always determined in individual confocal sections.

2.3.4 Stereological quantifications in spinal cord sections

Stereological counts (Coggeshall and Lekan, 1996) were performed in confocal image stacks of three randomly selected vibratome sections from the region up to 750 µm rostral to the lesion site and three sections from the region up to 750 µm caudal to the lesion site. Cell numbers were then calculated for the entire 1.5 mm surrounding the lesion site. Variability of values is given as standard error of the mean. Statistical significance was determined using the Mann-Whitney U-test ($p < 0.05$) or ANOVA with Bonferroni/Dunn post-hoc test for multiple comparisons.

2.3.5 Profile counts in spinal cord sections

$PCNA^+$ and $BrdU^+$ nuclear profiles in the ventricular zone (up to one cell diameter away from the ventricular surface) were determined in vibratome sections (50 µm thickness) in the same region of spinal cord. At least 6 sections were analysed per animal by fluorescence microscopy and values were expressed as profiles per 50 µm section. The observer was blinded to experimental treatments. Variability of values is given as standard error of the mean. Statistical significance was determined using the Mann-Whitney U-test (p < 0.05) or ANOVA with Bonferroni/Dunn post-hoc test for multiple comparisons.

2.3.6 Microinjection into zebrafish eggs

Freshly fertilized eggs were harvested 15 minutes after the light in the fish facility was switched on. Eggs were washed with autoclaved fishwater containing Methylene blue 10^{-5} % and arranged in a line in a petri dish containing 2% agarose in 1x PBS. To visualize the amount of injected liquid, 0.3 µl of 5% rhodamine dextran (MW = 10000) were added to a 1 µl aliquot of morpholino, mRNA, or Danieau solution. A glass micropipette (3 µm, GB 150F-8P, Science Products GmbH, Hofheim, D) was filled with the required solution by capillary force and attached to a micromanipulator (Microinjector Narishige, Intracel Ltd., Herts, UK). The solution was directly injected into the yolk of 1 - 4 cell staged eggs. Injected eggs were incubated in fishwater with Methylene blue at 28.5°C until the desired developmental stage was reached.

2.3.7 Whole mount immunohistochemistry

To detect proteins in 24 hpf zebrafish embryos, whole mount immunohistochemistry was performed. The chorions were removed and yolks were opened using an insect needle and fine forceps. Afterwards, embryos were fixed in 4% PFA containing 1% (v/v) DMSO for 45 min at RT. Then, embryos were washed in 1x PBS and incubated with blocking buffer to prevent unspecific binding of the primary antibody for 30 min at RT. Primary antibodies were diluted in blocking buffer and applied to the embryos and incubated overnight at 4°C. Three washing (1x PBS for 15 min) steps removed unbound primary antibody. To visualize primary antibodies, fluorescence- or HRP labelled secondary antibodies were diluted 1:200 in blocking buffer and applied to the embryos for 1h at RT. Unbound secondary antibody was removed by three washing steps with 1x PBS for 15 min each. To visualize the HRP signals, embryos were incubated in 0.5 mg/ml diaminobenzidine (DAB) in 1x PBS for 20 min at 4°C. The dark brown precipitate was developed by adding 1/10 volume of a 0.035% H_2O_2 solution in 1x PBS. After 5 - 10 min, the staining solution was removed, embryos were washed 3 times in 1x PBS and cleared in an ascending glycerol series (30, 50 and 70% glycerol in 1x PBS). Embryos were mounted in 70% glycerol.

2.3.8 Whole mount *in situ* hybridisation

To detect the expression patterns of mRNAs in 16-24 hpf zebrafish embryos, whole mount *in situ* hybridization was performed. Embryos at the desired developmental stages were anesthetized in 0.1% aminobenzoic acid ethyl methyl ester (MS222, Sigma-Aldrich, UK), dechorionated and fixed overnight in 4% PFA at 4°C. The following day, the embryos were washed 4 times with PBST (Phosphate Buffered Saline + 1% Tween®40) and incubated in 100% methanol (-20 °C) for 30 min. Methanol was removed using a descending methanol series (75 %, 50 % and 25 % methanol in PBST) and washed twice in PBST. To enhance penetration of the DIG labelled RNA probes, embryos were digested with 1.4 µg/ml recombinant Proteinase K (Roche, UK) in PBST for 10 min at RT. Two wash steps in 2 mg/ml glycine in PBST followed. Embryos were post-fixed in 4% PFA for 20 min at RT and subsequently washed 4 times with PBST to remove residual PFA. Embryos were prehybridized in hybridization buffer at 55°C for at least 3 hours. Hybridization with the DIG-labelled probes occurred at 55°C overnight. DIG-labelled probes were diluted 1:250- 1:4000 in hybridization buffer. After hybridization, embryos were washed twice in with 2x SSCT containing 50% formamide for 30 min, followed by a washing step in 2x SSCT for 15 min and two washing steps with 0.2x SSCT for 30 min. All washing steps were executed at 55°C. To prevent unspecific binding of the anti-DIG AP-conjugated antibodies, embryos were incubated for 30 min in 1% w/v Blocking Reagent (Roche, Mannheim, D) in PBST. Anti-Digoxigenin-AP antibodies (Roche, Mannheim, D) were diluted 1:2000 in Blocking Reagent and applied overnight at 4°C. To remove unbound antibody, embryos were washed 6 times in 1x PBST for 20 min. The washing solution was removed and the signal was developed in the dark with SIGMA FAST™ BCIP/NBT tablets (Sigma-Aldrich) until the reaction product became visible under a stereomicroscope. Sense probes were developed in parallel under the same conditions as the antisense probes and did not show any labeling. Finally, embryos were washed 3 times in 1x PBS and cleared in an ascending glycerol series (30, 50 and 70% glycerol in

1x PBS). The yolk sack was removed and embryos were mounted in 70% glycerol.

Whole mount hybridisation buffer:
5 ml deionized formamide
2.5 ml 20x SSC
10 µl Tween 20
100 µl 100 mg/ml yeast RNA (Sigma Aldrich, Deisenhofen, D)
2.38 ml DEPC-H_2O
10 µl 50 mg/ml heparin

2.4 Animal experiments

All fish are kept and bred in our laboratory fish facility according to standard methods and all experiments have been approved by the Home Office.

2.4.1 Perfusion fixation of adult zebrafish

After killing fish in 0.1% aminobenzoic acid ethylmethylester (MS222; Sigma, St. Louis, MO) they were transcardially perfused with 4% paraformaldehyde and post-fixed at 4°C overnight.

2.4.2 Spinal cord lesion of adult zebrafish

Before the spinal cord lesion, fish were kept for at least 24h in water with 1300 µS salt concentration to prevent bacterial or fungal infections. As described previously (Becker et al., 1997), fish were anesthetized by immersion in 0.033% aminobenzoic acid ethylmethylester in PBS for 5 min. A longitudinal incision was made at the side of the fish to expose the vertebral column. The spinal cord was completly transected under visual control 4 mm caudal to the brainstem-spinal cord junction. Afterwards the lesioned fish were kept in single tanks with high salt concentration and ESHA2000.

2.4.3 Retrograde tracing of adult zebrafish

Motor neurons in the spinal cord were retrogradely traced by bilateral application of biocytin to the muscle periphery at the level of the spinal lesion, as described previously (Becker et al., 2005),with the modification that biocytin was detected with Cy3-coupled streptavidin (Invitrogen) in spinal sections.

2.4.4 Intraperitoneal substance application

Animals were anaesthetised and intraperitoneally injected. We injected 5-bromo-2-deoxyuridine (BrdU, Sigma-Aldrich, UK) solution (2.5 mg/ml) at a volume of 50 µl at 0, 2, 4 days post-lesion. Analysis took place at 14 days post-lesion.

Cyclopamine was purchased from LC Laboratories (Woburn, MA, USA). The related control substance tomatidine (Sigma-Aldrich, UK). For intraperitoneal injections into adult fish cyclopamine and tomatidine were dissolved in HBC (45% (2-Hydroxypropyl)-beta-cyclodextrin) (Sigma-Aldrich, UK) and injected at a concentration of 0.2mg/ml in a volume of 25µl (equaling 10 mg/kg, Sanchez and Ruiz i Altaba, 2005) at 3, 6 and 9 days post-lesion. Analysis took place at 14 days post-lesion

3 RESULTS

3.1 Adult spinal cord regeneration

Zebrafish show functional regeneration after a lesion and the role of descending axons from the brainstem in this process has been studied extensively. Here we address the plastic changes occuring in the spinal cord. Specifically, I ask whether neurogenesis takes place in the lesioned spinal cord.

3.1.1 Spinal cytoarchitecture is not restored in a spinal lesion site

To determine in which spinal cord region neurons might regenerate, we analysed the overall organization of the regenerated spinal cord (Fig. 3 A) at 6 weeks post-lesion, when functional recovery plateaus. Electron-microscopic analysis, performed by Dr. Catherina G. Becker, indicates that both ends of the severed spinal cord fuse and form a thin tissue bridge which consists mainly of regenerated, partially re-myelinated, axons (Fig. 3 B). Therefore, substantial neurogenesis in the lesion site is unlikely.

The pre-lesioned spinal cord is still present after regeneration. Immediately adjacent to the axonal bridge the original cytoarchitecture is still found. Furthermore, white matter tracts are filled with myelin debris of degenerating fibres, indicating that this tissue was present before the lesion (Becker and Becker, 2001).

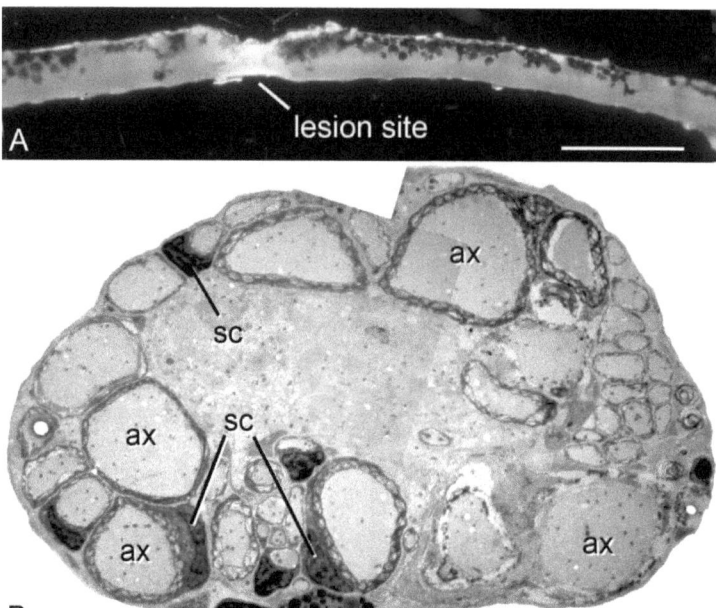

Fig. 3: The lesion site consists mainly of regenerated axons. **A:** A lateral stereo-microscopic view of a dissected spinal cord is shown (rostral is left). The dorsal aspect of the spinal cord is covered by melanocytes and the tissue bridging the lesion site appears translucent. **B:** An electron-microscopic cross-section through the lesion site is shown. The lesion site consists mainly of axons (ax), some of which are re-myelinated by Schwann cells (sc). Bar in A = 1 mm, in B = 5 µm.

3.1.2 A spinal lesion triggers ventricular proliferation

We analysed proliferation patterns in the lesioned spinal cord to determine in which region neuronal regeneration might take place. Proliferation activity in the spinal cord was studied by repeated injections of 5-bromo-2-deoxyuridine (BrdU). BrdU, a synthetic thymidine analogue, is incorporated into the DNA of dividing cells and later detected via immunohistochemistry. The injections were given at 0, 2 and 4 days post-lesion (dpl). Proliferation patterns were analysed at 2 weeks post lesion (wpl). In the unlesioned spinal cord only a few cells were labelled (Fig. 4, A left), indicating that cell division is a rare event. At 2 weeks post-lesion the number of newly generated cells in the spinal cord is significantly increased (p=0.0001, n = 3 animals) compared to the unlesioned situation (Fig. 4, middle, right). This increase is detectable up to 3.6 mm rostral and 3.6 mm caudal from the lesion site. Thus, it spans up to 1/3 of the entire spinal cord (Fig. 4, B). Numbers of newborn cells were highest close to the lesion site and around the central canal.

To determine the location of acutely proliferating cells we used an antibody recognizing the Proliferating Cell Nuclear Antigen (*PCNA*). In contrast to BrdU, which labels dividing cells permanently, the *PCNA* antibody only labels acutely proliferating cells in the early G1 and S phase of the cell cycle. This showed a significant increase in proliferating cells only in the ventricular zone. Already at 3 dpl the increase in ventricular proliferation was significant ($p < 0.0001$, n = 3 animals/time point) and peaked at 2 wpl. At 6 wpl, the proliferation was reduced again to levels that were not significantly different from those in unlesioned animals. This corresponds to functional recovery, which is complete at the same time point.

These findings suggest that, after a spinal lesion, new cells were primarily generated at the ventricle and then migrated out in to the parenchymal region.

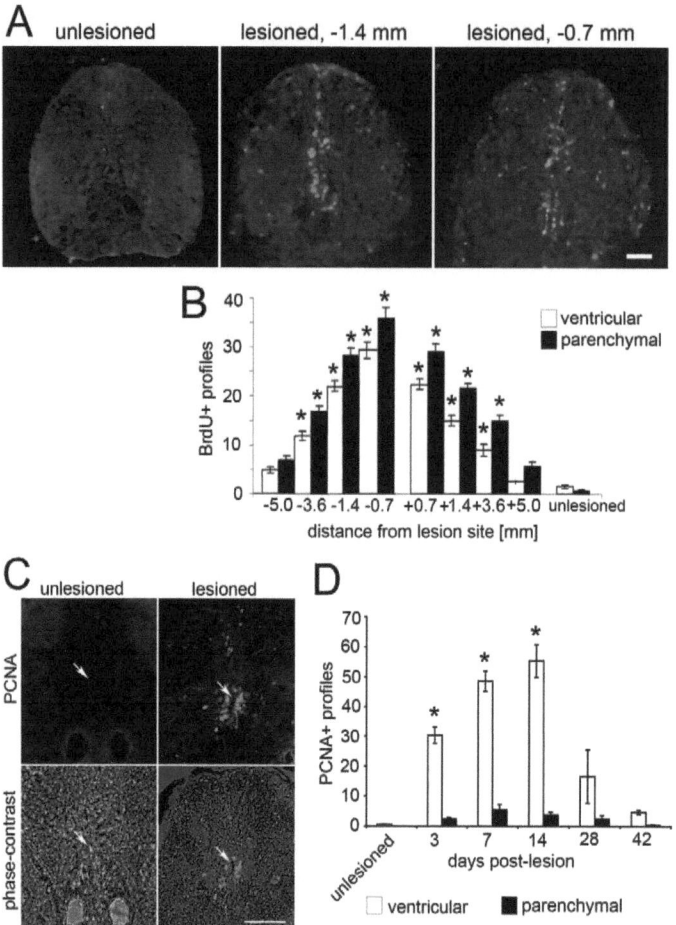

Fig. 4: Lesion-induced proliferation in the adult spinal cord. Confocal images of spinal cross-sections are shown (dorsal is up). **A:** BrdU labeling of spinal cross-sections shows a massive increase in labeling in the ventricular zone at 2 wpl (injections 0, 2, and 4 days post-lesion). The highest density of BrdU$^+$ cells is detectable in the ventricular zone close to the lesion site. **B:** Quantification of BrdU$^+$ profiles at 2wpl indicates significant proliferative activity up to 3.6 mm rostral and caudal to the lesion epicenter (n = 3 animals per treatment, p < 0.0001). **C:** PCNA immunohistochemistry indicates a strong increase in the number of proliferating cells in the ventricular zone (arrows) at 14 days post-lesion. **D:** The number of proliferating ventricular, but not

parenchymal cell profiles/section was significantly increased after a lesion and peaked at 2 wpl (n = 3 animals per time point, p <0.0001). Bar in A = 25 µm, in C = 50 µm.

3.1.3 Motor neurons show significant regenerative capacity

To determine whether neuronal death and/or regeneration occurs in the lesioned spinal cord, I focused on motor neurons, a cell type that never regenerates in mammals. To this end, numbers of GFP$^+$ motor neurons in *HB9:GFP* and *islet-1:GFP* transgenic animals were analysed (Higashijima et al., 2000; Flanagan-Steet et al., 2005). These lines express green fluorescent protein (GFP) under the control of the promotor for *HB9* or *islet-1*. In addition, antibodies against *islet-1/2*, *HB9* (also called *MNR2*) and transmitter synthesizing enzyme choline acetyltransferase (*ChAT*) proteins were utilised.

Islet-1/2 is a transcription factor of the LIM family and is expressed in various subpopulations of motor neurons in the spinal cord of adult zebrafish. The homeobox gene *HB9* is expressed in an overlapping population together with the *islet-1* and *islet-2* proteins as well as in *islet-1/2⁻* motor neurons (Renoncourt et al., 1998). The antibody against *ChAT* protein labels mature motor neurons.

3.1.3.1 Numbers of large and small motor neurons show dynamic changes after a lesion

Unlesioned *HB9:GFP* animals showed 132.5 ± 34.88 large GFP$^+$ motor neurons (diameter >12 µm, n = 4 animals) per 1500 µm spinal cord. Testing the specificity of the transgene with the corresponding antibody reveals that 97.8% (n = 3 animals) of the *HB9:GFP$^+$* cells were also *HB9* immunopositive. Of the large *HB9:GFP$^+$* cells, 80.6% (n = 3 animals) express choline acetyl transferase (*ChAT*), a marker of mature motor neurons. This indicated that most large GFP$^+$ cells were fully differentiated motor neurons. Furthermore, retrograde tracing from the muscle periphery in 8 weeks post-lesion animals with Biocytin, followed by detection with Streptavidin-Cy3, reveals that 52 of 55 traced cells were *HB9:GFP$^+$* (n=3). This indicates that large *HB9:GFP$^+$* cells are innervating muscle tissue and therefore are mature motor neurons.

The response of small (< 12 µm diameter) and large (> 12 µm diameter) *HB9:GFP$^+$* motor neurons to a lesion was determined (Fig. 5) in an area of 750 µm rostral and caudal to the lesion site. The number of large GFP⁺ cells was significantly reduced at 1 wpl (p = 0.0035, n = 4 vs. 3 animals) and 2 wpl (p= 0.0003, n = 4 vs. 11 animals). After 6 to 8 weeks the number of large motor neurons was increased again to levels that were not significantly different from the unlesioned situation (p = 0.0867, n = 4 unlesioned vs. 6 lesioned animals). This showed that the original number of mature *HB9⁺* motor neurons is decreased in response to the lesion event. Furthermore it indicates a trend in recovery of the number of large cells.

Numbers of small *HB9:GFP⁺* motor neurons responded inversely to the transection of the spinal cord. A significant increase after 2 wpl (p < 0.0001, n = 4 unlesioned vs. 11 lesioned animals) was followed by a significant decrease in number of small neurons at 6 to 8 weeks (p = 0.0002, n = 11 animals at 2 wpl vs. 6 animals at 6 to 8 wpl).

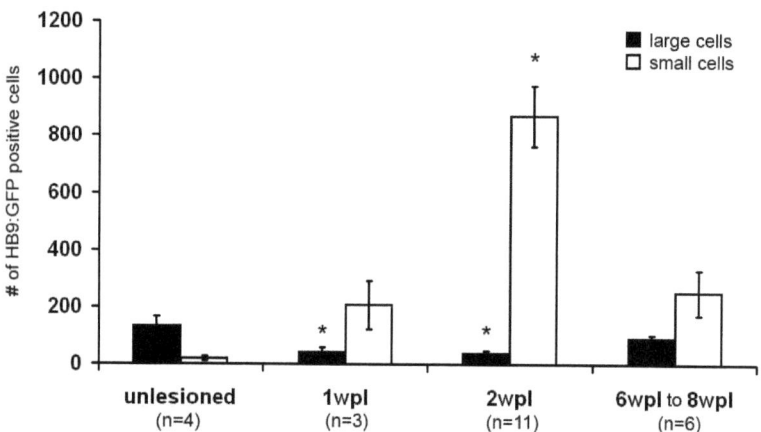

Fig 5: Dynamic changes in the numbers of *HB9:GFP⁺* motor neurons after a lesion. A spinal cord lesion induces an increase in the number of small and a decrease in the number of large motor neurons at 2 wpl. At 6 to 8 wpl, the population of large motor neurons partly recovers, while numbers of small *HB9:GFP⁺* cells return to original levels. Stereological counts of *HB9:GFP⁺* cells calculated to 1500µm around the lesion site are given.

This transient, more than 43-fold increase, in the number of small $HB9:GFP^+$ motor neurons indicates a highly dynamic response in the number of spinal motor neurons to the lesion event. In addition, the time course matches that of the functional recovery, indicating a possible link between motor neuron regeneration and functional recovery.

Using the *islet-1/2* antibody the spatial distribution of differentiating motor neurons was analysed in 14 µm cryosections at increasing distances from the lesion site (Fig. 6). Close to the lesion site (0-250 µm) the number of islet-1/2 positive cell profile counts is highest and significantly increased 2 wpl (p = 0.0253, n = 5 animals each group) compared to unlesioned controls. This corresponds to proliferative activity in the ventricular zone, which is also highest close to the lesion site (Fig. 4 D).

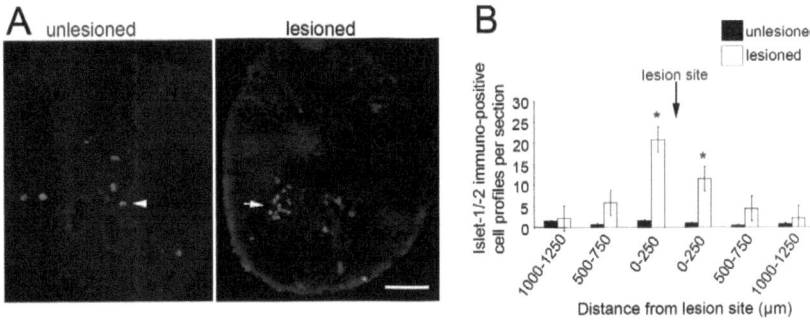

Fig 6: *Islet-1/-2* immunohistochemistry confirms an increase in the number of differentiating motor neurons. **A:** Few large nuclei (arrowhead) are visible in the unlesioned spinal cord. In the lesioned situation, clusters of small *islet-1/-2* immunopositive cell nuclei appear in the ventro-lateral spinal cord (arrow). **B:** Numbers of *islet-1/-2* immunopositive cell profiles were determined in cryosections (14 µm in thickness) for the regions indicated, showing a significant increase in *islet-1/-2* immunopositive cell profiles around the lesion site. (n = 5 unlesioned animals; n = 5 animals at 2 wpl; p = 0.0253). Bar = 50 µm.

3.1.3.2 Small motor neurons are newly generated after a lesion

To directly address whether motor neurons were newly generated, BrdU was injected into *HB9:GFP* and *islet-1:GFP* transgenic animals at 0, 2, and 4 dpl post-lesion and the number of double labelled neurons was determined at 2 wpl.

At 2 wpl there was an increase in the number of small *islet-1:GFP* positive cells, which was statistically significant compared with the unlesioned situation (unlesioned: 27 ± 3.9 cells, n = 5 animals, 2 wpl: 870 ± 244.9 cells, n = 4 animals, p = 0.0139). In BrdU injected animals, 184 ± 49.3 small cells (n = 3 animals, p = 0.0104) were double labelled with the transgene and BrdU immunohistochemistry. In the unlesioned controls no double-labelled cells (n = 5 animals) were found (Fig. 7).

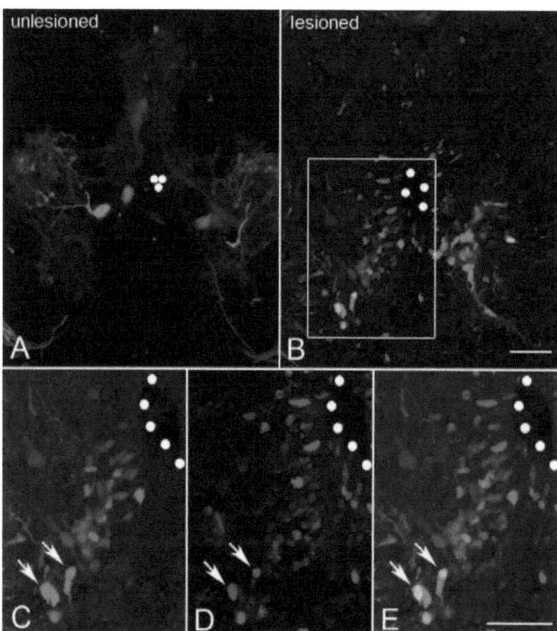

Fig 7: Newly generated small *islet-1:GFP*[+] cells in the lesioned spinal cord. Cross-sections through the spinal cord of unlesioned **A:** and lesioned **B-E:** animals at 2 weeks post-lesion are shown. In unlesioned animals only large GFP[+] cells are detectable, whereas many smaller GFP[+] cells are present in the ventrolateral aspect of the lesioned spinal cord. Many of these cells are also BrdU[+], as indicated by arrows in

the higher magnification **C-E:** of the area boxed in B. Dots outline the ventricle. Bars = 25 μm.

The *HB9:GFP* transgenic fish confirms these observations: at 2 wpl the small *HB9:GFP$^+$* cells were increased from 20.0 ± 7.66 in the unlesioned situation (n = 4 animals) to 869.5 ± 106.78 (n = 11 animals, p < 0.0001). In this transgenic fish, 200.0 ± 46.2 cells (n = 7 animals, p = 0.0076) were double-labelled by the transgene and BrdU at 2 wpl (Fig. 8). In the unlesioned spinal cord only one double-labelled motor neuron was observed (n = 4 animals). Even a BrdU injection protocol extended to the maximum number of injections tolerated by the fish (injections at 0, 2, 4, 6, 8 days post-lesion, analysis at 14 days post-lesion) did not yield any *HB9:GFP$^+$*/BrdU$^+$ cells in unlesioned fish (n = 5 animals). Letting the fish swim in BrdU-treated water in order to label all newly generated cells over the entire time of the experiment does not show sufficient labeling of dividing cells (Dr. Thomas Becker, personal communication). Hence the unlesioned mature spinal cord appears virtually quiescent with respect to motor neuron generation. However, low rates of motor neuron formation may have been missed due to the limited metabolic availability of BrdU.

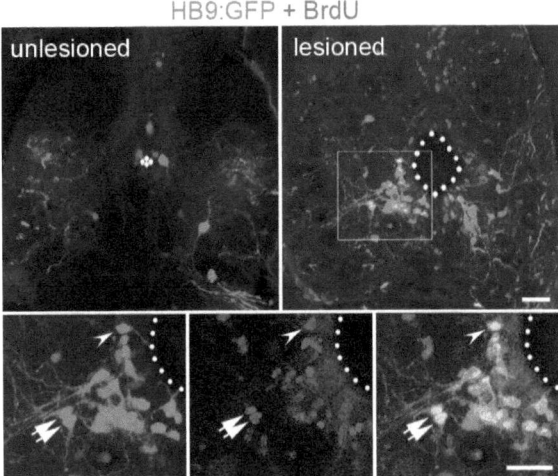

Fig. 8: Generation of new motor neurons in the lesioned spinal cord. *HB9:GFP*/BrdU double-labelled neurons are present in the lesioned, but not the unlesioned, ventro-

lateral spinal cord. These cells (boxed in upper right and shown in higher magnification in bottom row) bear elaborate processes (arrows) or show ventricular contact (arrowhead). Bars = 25 µm.

3.1.3.3 Lesion induces cell death

The number of large $HB9^+$ motor neurons decreases significantly after a lesion. We performed TUNEL staining in *HB9:GFP* transgenic fish at 3 dpl and found TUNEL$^+$/*HB9:GFP*$^+$ cells (Fig. 9). The apoptosis marker TUNEL labels the nuclei of cells undergoing programmed cell death (Hewitson et al., 2006).

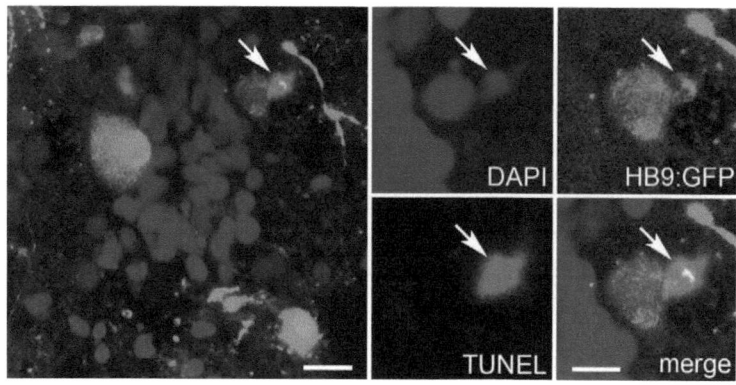

Fig. 9: Lesion induced apoptosis at 3 dpl. *HB9:GFP* (green), DAPI nuclear staining (blue) and TUNEL staining (red). Triple labelling indicates apoptotic motor neurons (arrow). Bars: left 15µm, right 8 µm.

3.1.3.4 Different subpopulations of newly generated motor neurons may be present in the lesioned spinal cord

The *islet-1:GFP* and the *HB9:GFP* transgenic animals show a similar distribution of small motor neurons in the ventral horn of the lesioned spinal cord. For *islet-1*, the transgene expression confirms the expression of the endogenous gene because 89.5 % of the *islet-1:GFP*$^+$ cells were *islet-1/2* immunopositive at 2 wpl. The small proportion of cells only labelled by GFP in *islet-1:GFP* animals may result from higher stability of the GFP than endogenous islet-1 detected by the antibody. In contrast, a substantial proportion, 51.7 %, of *HB9:GFP*$^+$ cells were not double-labelled by the *islet-1/2* antibody and many cells were exclusively labelled by the islet-1/-2 antibody in both transgenic lines (55.7 % in the *HB9:GFP* and 35.4 % in the *islet-1:GFP* fish) (Fig. 10). This suggests heterogeneity among newly generated motor neurons with respect to marker expression (William et al., 2003).

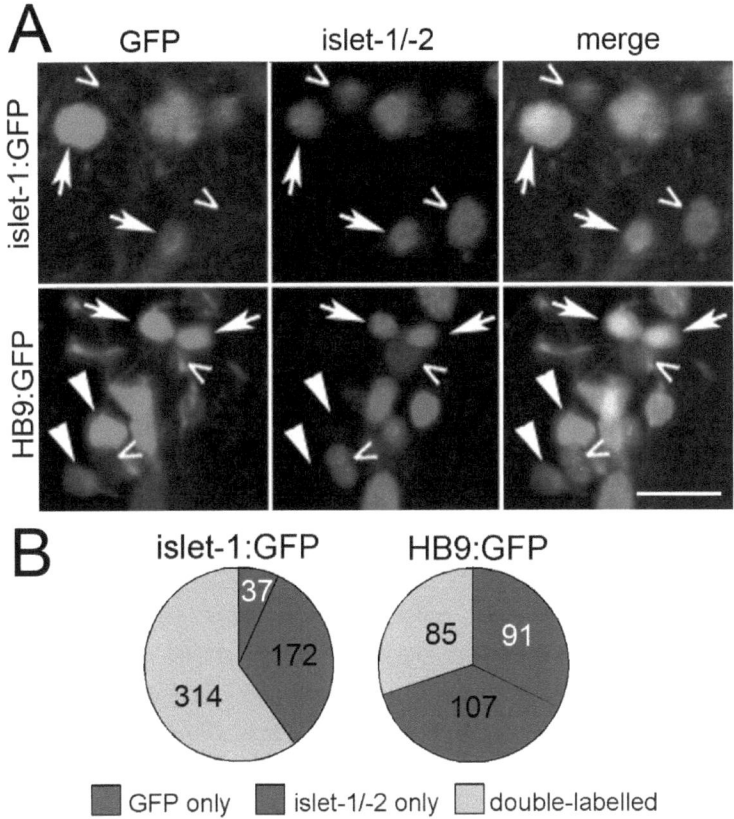

Fig. 10: Partial overlap of *islet-1/-2* immunohistochemistry and transgenic motor neuron markers in the lesioned spinal cord. A: *Islet-1:GFP*[+] cells are double-labelled by the *islet-1/-2* antibody, confirming specificity of transgene expression. A substantial proportion of *HB9:GFP*[+] cells are not double-labelled by the antibody and many cells are only labelled by the *islet-1/-2* antibody in both transgenic lines, suggesting that different types of cells were generated after a lesion. Arrows indicate double-labelled neurons, arrowheads indicate neurons only labelled by the transgene and open arrowheads point to cells only labelled by the antibody. B: Summations of all cells counted in six sections (50 μm thickness) per animal from the region of 1.5 mm

surrounding the lesion site (n = 3 animals for each transgene) are indicated. Bar = 25 µm.

3.1.3.5 Newly generated motor neurons show terminal differentiation and may be integrated into the spinal network

To determine whether newly generated motor neurons fully matured, expression of *ChAT*, a marker for terminally differentiated motor neurons (Arvidsson et al., 1997), and coverage of motor neurons by $SV2^+$ contacts, a marker for synaptic coverage, was analysed. In the unlesioned situation 80.6% (n = 3 animals) of the large $HB9:GFP^+$ cells expressed *ChAT*, indicating that the majority of $HB9:GFP^+$ cells were mature motor neurons. Small $HB9:GFP^+$ cells were rarely found. Furthermore, all $ChAT^+$ cells were covered with synapses in the unlesioned spinal cord.

At 2 wpl, small $HB9:GFP^+$ neurons were rarely *ChAT* positive (2.8%, n = 3 animals) and did not receive $SV2^+$ contacts (Fig. 11, upper row). Of the large $HB9:GFP^+$ cells, 36.4% (n = 3 animals) were double labelled with *ChAT* and often not covered with *SV2* labelled synapses (Fig. 11, middle row). This indicates that most small and some of the large $HB9:GFP^+$ neurons were immature at 2 wpl.

To determine whether newly generated motor neurons show terminal differentiation and network integration at later stages of regeneration, BrdU injections at day 0, 2 and 4 were combined with anti-*ChAT* and anti-*SV2* immunohistochemistry. At 6 wpl 29.3 ± 23.14 $ChAT^+$ cells/1500µm (n = 3 animals) were also $BrdU^+$ and extensively covered with *SV2* labelled synapses. The inset in the lower row indicates that similar cells are part of the typical cytoarchitecture of the unlesioned spinal cord (Fig.11, lower row). These observations are consistent with the hypothesis that newly generated motor neurons can fully mature and integrate into the spinal network.

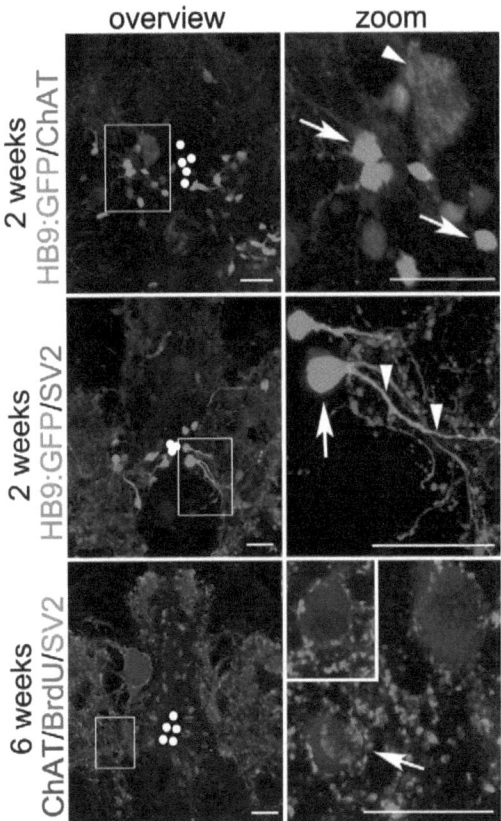

Fig. 11: Maturation of newly generated motor neurons. Confocal images of spinal cross-sections are shown (dorsal is up). Clusters of newly generated *HB9:GFP*[+] motor neurons are *ChAT* (arrow in top row indicates a *ChAT*[+]/*HB9:GFP* differentiated motor neuron). Somata (arrow in middle row) and proximal dendrites (arrowheads in middle row) receive few *SV2*[+] contacts at 2 wpl. At 6 wpl, *ChAT*[+]/BrdU[+] somata are decorated with *SV2*[+] contacts (arrow in bottom row), inset: unlesioned situation. Bars = 25 μm.

3.1.3.6 Evidence for motor axon growth out of the spinal cord

To determine whether newly generated motor neurons grow axons out of the spinal cord, we applied the retrograde neuronal tracer biocytin to the muscle tissue surrounding the lesion site of BrdU injected animals. Biocytin tracing marks muscle-innervating neurons in the spinal cord, which are bona fide motor neurons. Fish were injected at 12, 13 and 14 dpl with BrdU and biocytin was applied at 42 dpl. Out of 4 fish, we found one BrdU$^+$/biocytin$^+$ cell (Fig.12), indicating that this newly generated cell extended an axon out of the spinal cord. The ventro-lateral position of the cell in the spinal cord is consistent with a motor neuron identity of this cell. One reason for the scarcity of these double-labelled cells may be that BrdU labels only a sub-population of newly generated motor neurons (approximately 25% at 14 dpl) and retrograde tracing does not label all motor neurons, such that overlap of the two markers may be a rare event. However, this observation suggests that newly generated motor neurons are capable of regenerating a peripheral axon.

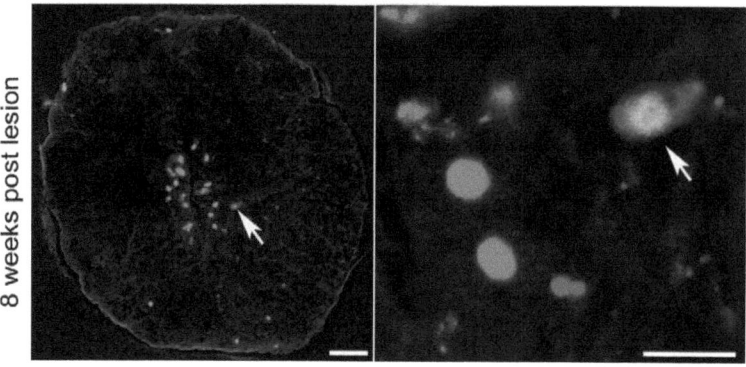

Fig.12: Retrograde tracing of a newborn motor neuron in the spinal cord from the muscle periphery. Confocal images of the same spinal cross-section are shown in low (left) and high (right) magnification (dorsal is up). Arrows point to the same biocytin/BrdU doubled labelled motor neuron at 8wpl. Bars = 50 µm (left), 15 µm (right).

3.1.4 *Olig2:GFP$^+$* ependymo-radial glial cells are potential motor neuron stem cells in the adult spinal cord

Olig2 expressing cells give rise to motor neurons during development. To determine whether this cell type also exists during adult regeneration and plays a similar role, we investigated adult expression of *olig2*. A transgenic fish expressing GFP under the control of the *olig2* promoter outlines the entire morphology of the ventricle-contacting *olig2* positive cells, including long radial processes in unlesioned animals (Fig. 13, upper row). Additionally, the transgene marks *olig2*-expressing oligodendrocytes. These cells are morphologically distinguishable from *olig2⁺* ependymo-radial glial cells and are distributed in the parenchyma of the spinal cord. For this study, I focused on the ependymo-radial glial cells because they are similar in morphology to radial glial progenitor cells in the developing CNS and their somata are located at the ventricle, where lesion-induced proliferation takes place. Indeed, *olig2:GFP⁺* ependymo-radial glial cells respond to a spinal cord transection with proliferation, as demonstrated by immunohistochemistry for *PCNA*, which marks acutely proliferating cells (Fig. 13, lower row).

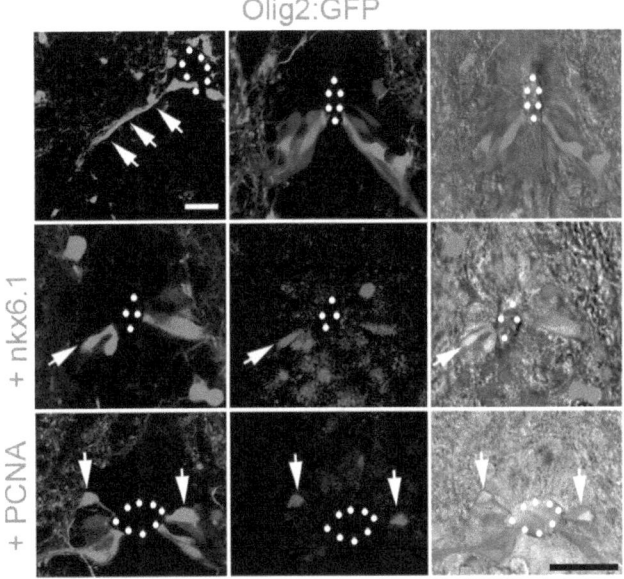

Fig. 13: *olig2:GFP+* cells have long radial processes (arrows in upper left), contact the ventricle (upper row, middle and right), and are double-labelled (arrows) with *nkx6.1* and *PCNA* antibodies at 2 wpl. Confocal images of cross-sections are shown. Dots outline the ventricle. Bar = 25 µm.

To determine whether other transcription factors, known to be important for motor neuron development, are also expressed in *olig2:GFP⁺* ependymo-radila glial cells during regeneration, we double labelled with antibodies against *nkx6.1*. Developmentally, the homeodomain transcription factor *nkx6.1* is part of the mechanisms regulating *olig2* expression. It promotes *olig2* expression at an early stage in development and represses it at a later stage in chicken (Liu et al., 2003). In the adult spinal cord of zebrafish, *olig2:GFP*$^+$ ependymo-radial glial cells expressed *nkx6.1* at 2 wpl (Fig. 13, middle row). The presence of *nkx6.1* in *olig2*$^+$ ependymo-radial glial cells after a lesion indicates that during regeneration of motor neurons a gene expression program similar to development could occur.

3.1.4.1 Lineage tracing indicates that *olig2:GFP*$^+$ ependymo-radial glial cells are motor neuron progenitor cells.

To directly demonstrate that *olig2*$^+$ ependymo-radial glial cells are the progenitor pool for new motor neurons, I used GFP expression as a stable marker for linage tracing. *Olig2*$^+$ ependymo-radial glial cells that give rise to motor neurons may still be GFP$^+$ when starting to express the motor neuron specific markers *HB9* and *islet-1*. This is because GFP is a relatively stable protein (Tallafuss and Bally-Cuif, 2003).

Comparing unlesioned and lesioned spinal cord of *olig2:GFP* transgenic fish with anti-*HB9* immunohistochemistry reveals that after a lesion *olig2*$^+$ ventricular cells differentiate to *HB9* expressing motor neurons (Fig.14, middle row). In the unlesioned fish no *GFP*$^+$/*HB9*$^+$ cells could be detected (Fig. 14, upper row) versus 204.0 ± 32.29 *GFP*$^+$/*HB9*$^+$ cells per 1500 µm around the lesion site in the group of the lesioned animals (n = 3 animals per group).

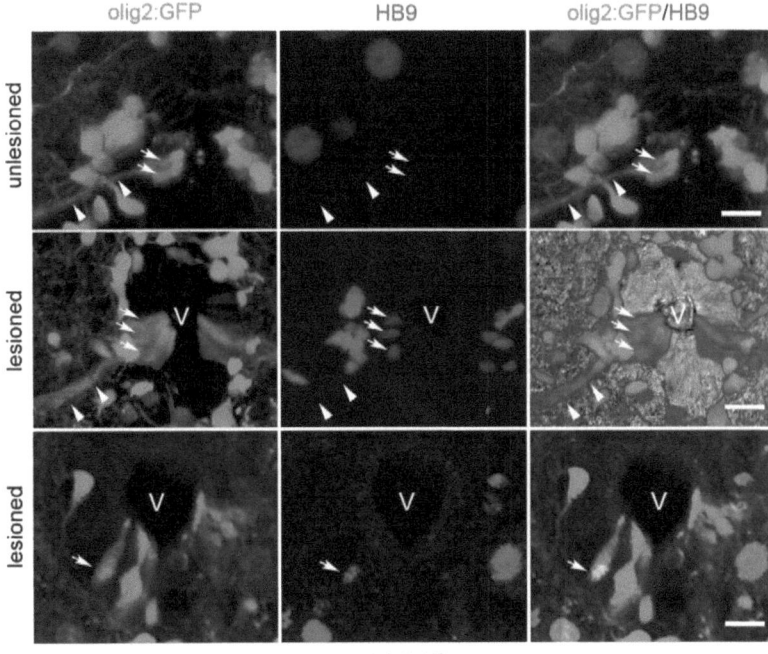

Fig. 14: Confocal images of spinal cross-sections unlesioned vs. 2 wpl are shown (dorsal is up). *Olig2:GFP*[+] progenitor cells (arrows) have long radial processes (arrowheads), contact the ventricle (outlined by dots), and are double-labelled with *HB9* or *islet-1/2* antibodies at 2 wpl, but not in the unlesioned spinal cord. Bars in A = 25 µm; Bars in B = 7.5 µm (upper row), 15 µm (middle and lower row).

Similarly, transgenic zebrafish showed ependymo-radial glial cells double-labelled for *olig2:GFP* and *islet-1/-2*[+] cells at 2 wpl (Fig. 14, lower row). In the 1500 µm around the lesion site, 34.3 ± 8.93 cells (n = 4 animals) were double labelled. This indicates that *olig2*[+] ependymo-radial glial cells have the capacity to proliferate and to give rise to cells expressing markers for motor neurons in response to a lesion. Moreover, this indicates a molecular switch of the *olig2* expressing ependymo-radial glial cells from a gliogenic to a motor neuron cell fate after a lesion.

3.1.4.2 *Olig2:GFP+* ependymo-radial glial cells are polysialic acid and GFAP negative

Polysialic acid (PSA) and GFAP are progenitor cell markers (Rutishauser, 2008) and (Ninkovic and Götz, 2007). Surprisingly, *olig2:GFP$^+$* ependymo-radial glial cells are selectively GFAP (data not shown) and PSA immuno-negative in unlesioned animals (Fig. 15). Other ependymo-radial glial cells around the entire ventricle express these antigens. However, *olig2:GFP$^+$* ependymo-radial glial cells express another progenitor cell marker, brain lipid binding protein (Park et al., 2007). This indicates that *olig2:GFP$^+$* ependymo-radial glial cells express a unique set of progenitor cell markers.

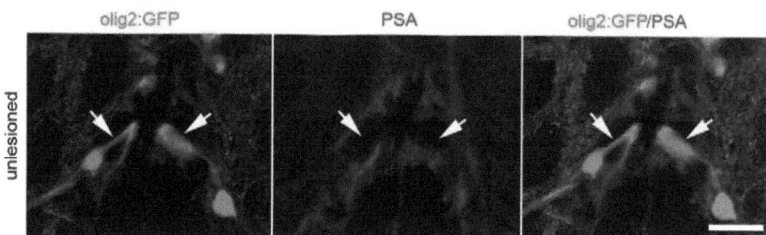

Fig 15: Anti-PSA immunohistochemistry in unlesioned fish. *Olig2:GFP$^+$* ependymo-radial glial cells (arrows) are PSA$^-$. Cross sections of the spinal cord at the ventricular zone are shown. Bar = 10 µm.

3.1.4.3 Ependymo-radial glial cells, including *olig2:GFP$^+$* cells, in the spinal cord are label-retaining cells

To determine whether ventricular cells have stem cell characteristics, I tested for BrdU retention over an extended time period, an indicator of slow proliferation (Chapouton et al., 2006). Lesioned animals were injected with a single pulse of BrdU at 14 days post-lesion and the number of ventricular BrdU$^+$ cells was assessed around the entire ventricle at 4 hours and 14 days post-injection (Fig.16). There was no significant difference in the number of BrdU labelled cells at both time points ($p = 0.7237$), indicating the presence of slowly proliferating cells in the ventricular zone.

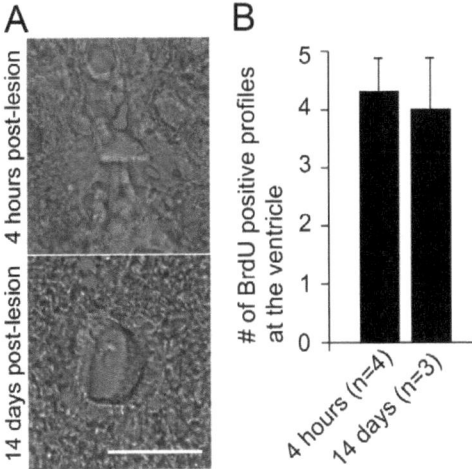

Fig 16: The ventricular zone contains label-retaining cells. **A:** After a single injection of BrdU at 2 wpl and subsequent immunohistochemical analysis at 4 hours or 14 days after injection, BrdU labelled cell profiles were found at the ventricle. **B:** Numbers of BrdU$^+$ cell profiles around the ventricle (up to two nuclei away from the ventricular surface) were determined in confocal image stacks of three randomly selected vibratome sections each from the region up to 750 µm rostral and caudal to the lesion site. This analysis indicates no significant differences in the number of BrdU labelled cells at both time points (p = 0.7237). Bar = 50 µm.

To examine whether the *olig2:GFP$^+$* ventricular zone also contained label-retaining cells, I repeated the same experiment in *olig2:GFP* transgenic animals. Numbers of *olig2:GFP$^+$*/BrdU$^+$ ependymo-radial glial cells were not significantly different between the two time points (4 hours: 60 ± 11.5 cells, n = 5 animals; 14 days: 53 ± 13.3 cells, n = 4 animals, p = 0.6). This indicates that *olig2:GFP$^+$* cells did indeed retain label (Fig. 17).

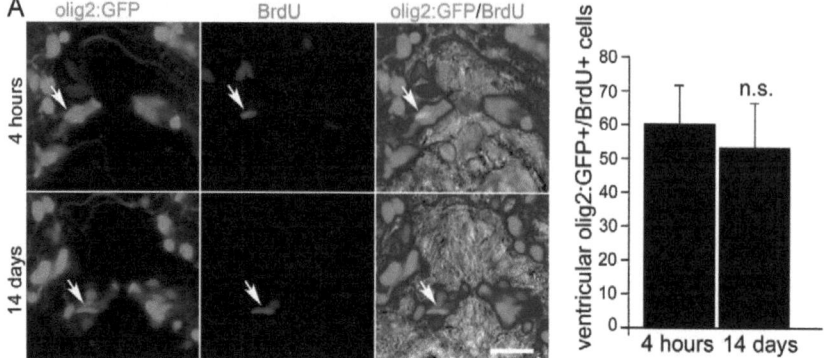

Fig. 17: Label retention in *olig2:GFP* ependymo-radial glial cells. **A:** A subpopulation of *olig2:GFP⁺* ependymo-radial glial cells is BrdU⁺ at 4 hours and 14 days after a single application of BrdU at 2 wpl. Bar = 15 µm. **B:** No significant differences in the number of *olig2:GFP⁺*/BrdU⁺ cells were observed between both time points of analysis.

Both label-retention experiments indicate the stem cell characteristics of ventricular cells, specifically of *olig2⁺* ependymo-radial glial cells.

Thus far I was able to show that in response to spinal cord lesion new motor neurons were generated. Some of these small motor neurons matured and integrated in the intraspinal circuitry. Evidence suggests that the origin of these new motor neurons is the *olig2⁺* ependymo-radial glial cell domain. These ventricular cells even possess stem cell characteristics.

3.1.5 Expression of ventral neural tube markers is increased in a developmentally appropriate pattern

To determine the signals that may induce motor neuron regeneration we analysed expression of *sonic hedgehog* (*shh*), a key player in the organisation of spinal cord patterning in development (Lewis and Eisen, 2001). Transgenic zebrafish, expressing green fluorescent protein (GFP) under the promotor of *shh* revealed an increase in *shh* expression in response to a spinal cord lesion at 2 wpl (Fig. 18, left column). These cells form the very ventral region of the ventricle. In situ hybridisation with a probe against *shh* mRNA confirms the localization of the transcript and the upregulation in response to a spinal cord transection.

Olig2 is a downstream gene of *shh* signalling. In situ hybridisation showed a lesion induced upregulation of *olig2* expression at the enlarged ventricle, conterminous to the *shh* domain (Fig. 18, right column), suggesting that increased *shh* expression induced increased *olig2* expression in the neighbouring ventricle zone.

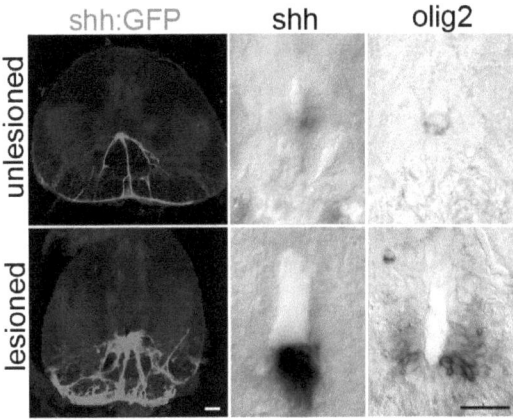

Fig. 18: Lesion-induced expression of *shh* and *olig2* at the ventricle of the lesioned spinal cord. *Shh:GFP* and in situ hybridisation signals are increased at 2 wpl. Cross-sections are shown. Bars = 25 µm.

Combination of the *shh:GFP* transgenic animals with immunohistochemistry detecting *PCNA* reveals that the ventricular *shh:GFP*[+] cells are proliferating at 2 weeks post-lesion (Fig. 19). This finding is consistent with the increase in the density of *shh:GFP*[+] cells and the intensity and region of *shh* RNA expression (Fig. 18, middle column).

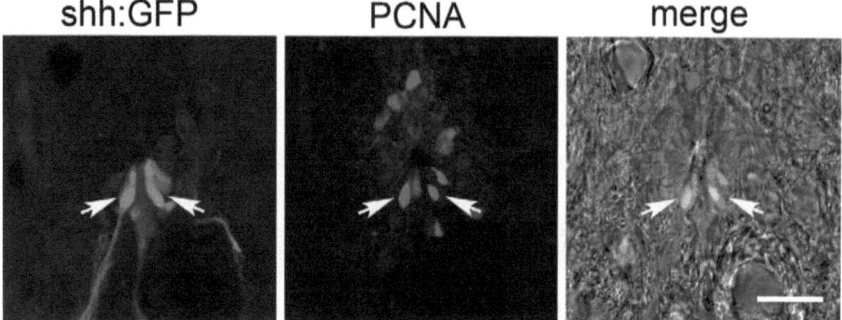

Fig. 19: *Shh:GFP*⁺ ependymo-radial glial cells proliferate after a lesion. GFP⁺ cells have a radial morphology, contact the ventricle at its ventral edge and are labelled by a *PCNA* antibody (arrows) at 2 wpl. Bar = 20 μm.

The increase in the expression of *shh* after a lesion and the increase of *olig2* expression in the adjacent *olig2* domain is consistent with the assumption that *shh* could be a key player in motor neuron regeneration.

Further important transcription factors that determine the cell fate of progenitor cells in the developing spinal cord are the ventrally expressed *nkx6.1* in combination with the *olig2:GFP* expression and the medio-dorsally expressed *pax6* (Becker and Becker, 2007).

Both *nkx6.1* and *pax6* were upregulated after a spinal cord lesion in the adult fish. Low levels of *nkx6.1* observed in the unlesioned ventral spinal cord were increased at 2 weeks post-lesion, but expression was still restricted to the ventral ventricular zone (Fig. 20, upper rows). At 1 wpl *pax6* was expressed in the lateral and dorsal domain of the ventricular zone and was upregulated in response to a spinal cord lesion in the same domain (Fig. 20). Double labelling of *nkx6.1* in *olig2:GFP* transgenic fish (Fig. 21) indicates that the *olig2:GFP*⁺ ventricular zone is included in the *nkx6.1*⁺ zone, with the *nkx6.1*⁺ zone extending slightly more dorsally than the *olig2:GFP*⁺ zone. This is comparable to the spatial relationships of the two expression domains in the developing neural tube.

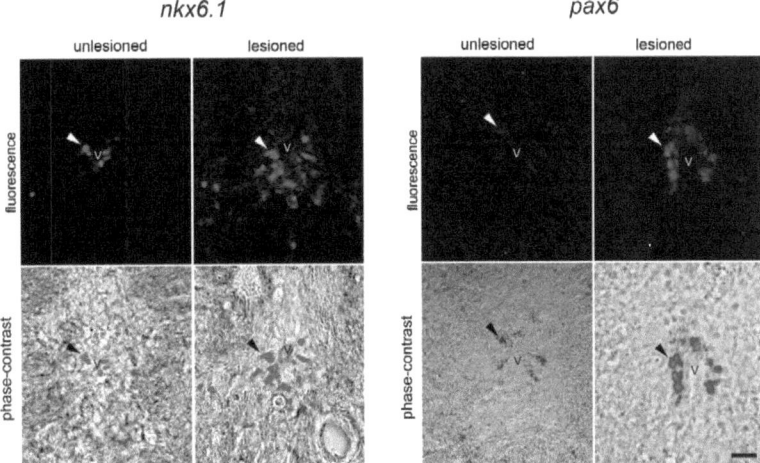

Fig 20: *Nkx6.1* and *pax6* expression is increased in the lesioned spinal cord. Labeling intensity of *nkx6.1* is increased around the ventral ventricle (arrows) at 2 wpl. *Pax6* immunohistochemistry shows upregulation of *pax6* after spinal cord lesion in the upper ventral and dorsal part of the spinal cord. 14 µm cryosections, unlesioned (overexposed to show low levels of immunoreactivity) vs. 1 wpl. Bar 20 µm.

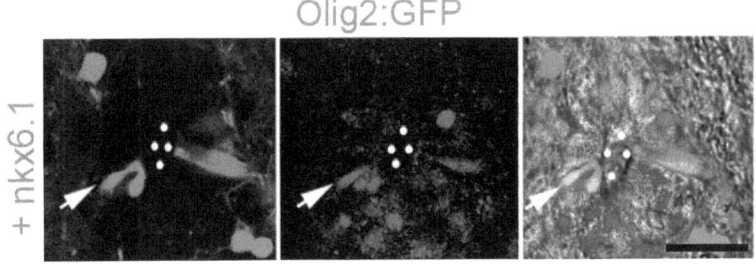

Fig. 21: *olig2:GFP+* ependymo radial glial cells are *nkx6.1* immunopositive in the lesioned adult spinal cord. Double-labelled (arrows) with *nkx6.1* antibody at 2 wpl. Confocal images of cross-sections are shown. Dots outline the ventricle. Bar = 25 µm.

These data indicate that transcription factors are expressed in different dorso-ventral domains of the neural tube. Additionally, the patterns in the ventricular zone of the unlesioned adult spinal cord are comparable to the lesioned situation, but the expression is increased after a spinal lesion.

3.1.6 Cyclopamine inhibits *shh* dependent motor neuron regeneration

If increased *shh* expression is involved in the generation of new motor neurons after a lesion, a pharmacoligical block of this signaling pathway should reduce the number of newly generated motor neurons. To test this hypothesis, the *shh* signal was experimentally reduced by injections of cyclopamine, a specific small molecule inhibitor of *shh* signalling (Park et al., 2004).

To control the specific activity of cyclopamine I incubated embryos with the substance. This treatment resulted in cyclopia and loss of motor axons (Tab.1), which is consistent with published actions of cyclopamine in zebrafish (Park et al., 2004) . In vehicle treated controls none of the animals showed cyclopia.

Tab. 1: Cyclopamine activity. 24 hpf *HB9:GFP* embryos were analysed after Cyclopamine treatment (n = 6 animals/group were analysed).

	treatment		vehicle control	
	axons/fish	Cyclopia	axons/fish	Cyclopia
100 µM Cyclopamine/EtOH	1.2 ± 0.54	7 of 15	46.3 ± 0.95	0 of 15
50 µM Cyclopamine/EtOH	2.3 ± 0.61	5 of 15	46.3 ± 1.09	0 of 15
5 µM Cyclopamine/EtOH	9.7 ± 2.35	1 of 14	48.0 ± 0.52	0 of 14

The loss of motor axons was analysed in the transgenic line *HB9:GFP* after cycopamine incubation from 6 to 24 hpf. Here the primary motor axons were massively effected. For *islet-1:GFP* embryos incubated from 24 to 72hpf the loss of motor neurons in the spinal cord is clearly visible (Fig. 22). This indicates that the *shh* pathway is necessary for the formation and development of motor neurons in development.

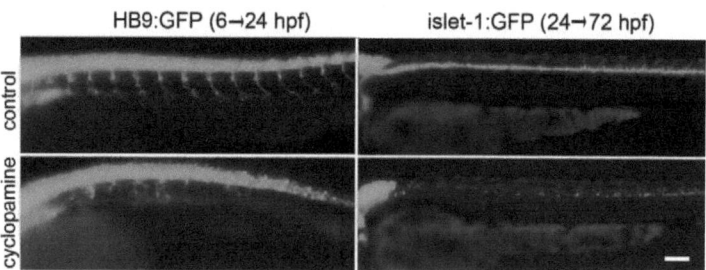

Fig. 22: Specific activity of Cyclopamine in transgenic lines *HB9:GFP* and *islet-1:GFP*. Control animals show normal axon growth, 5 µM Cyclopamine show severe effect on

motor axon outgrowth (*HB9:GFP*) and loss of motor neurons (*islet-1:GFP*). Rostral is left, Bar = 25 µm.

Adult spinal cord lesioned zebrafish were repeatedly injected with the *sonic hedgehog* inhibitor cyclopamine at 10 mg/kg, a mouse specific non toxic concentration (Ecke et al., 2008), 3, 6 and 9 days post operation. To avoid the toxic effects of ethanol, I used (2-Hydroxypropyl)-β-cyclodextrin (HBC) as solvent. This solvent has negligible effects on the activity of cyclopamine (Tab.2).

Tab. 2: Cyclopamine activity in HBC at 24 hpf. *HB9:GFP* embryos were analysed after Cyclopamine treatment (n = 6 animals/group were analysed).

	treatment		vehicle control	
	axons/fish	Cyclopia	axons/fish	Cyclopia
100 µM Cyclopamine/ HBC	3.7 ± 0.88	6 of 15	47.7 ± 0.95	0 of 15
50 µM Cyclopamine/ HBC	6.0 ± 1.91	4 of 15	46.7 ± 1.33	0 of 12
5 µM Cyclopamine/ HBC	30.8 ± 2.15	0 of 14	46.0 ± 1.15	0 of 15

Injecting cyclopamine into spinal-lesioned adult animals highly significantly reduced the number of newly generated motor neurons within 1.5 mm surrounding the lesion site (377 ± 45.7 cells; n = 9 animals) compared with animals injected with the related but ineffective substance tomatidine (747 ± 42.2 cells; n = 10 animals; p = 0.0004) at 2 wpl (Fig. 23 A, upper row).

Moreover, *shh* is a mitogen (Fuccillo et al., 2006) and I tested whether cyclopamine reduces ventricular proliferation in this region of the lesioned spinal cord by determining the numbers of *PCNA*[+] ventricular cells after the same cyclopamine treatment scheme. The number of *PCNA*-labelled cell profiles at the ventricle in cyclopamine-injected animals (45 ± 2.8 profiles/section; n =16 animals) was significantly lower compared with tomatidine-injected control animals (60 ± 7.0 profiles/section; n = 10 animals; p = 0.027; one-tailed test; Fig. 23 A, lower row). Thus, *shh* signalling appears to play a role for progenitor cell proliferation and motor neuron differentiation.

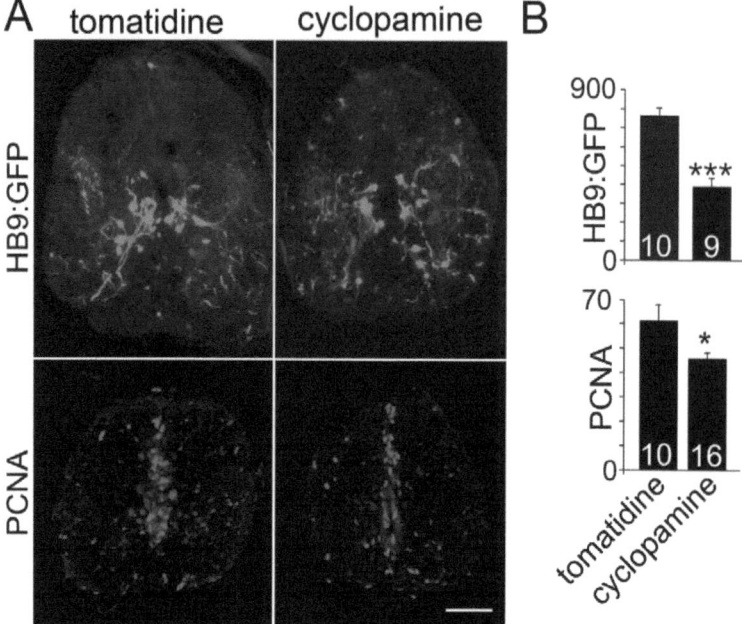

Fig 23: Cyclopamine treatment reduces the number of differentiating motor neurons and ventricular proliferation. **A:** Spinal cross-sections show reduced numbers of *HB9:GFP*$^+$ cells and *PCNA*$^+$ cells after cyclopamine treatment at 2 wpl. **B:** Numbers of *HB9:GFP*$^+$ cells and *PCNA*$^+$ profiles are significantly reduced in cyclopamine injected animals. Numbers of animals/treatment are given. Bar = 50 µm.

To assess whether cyclopamine injections specifically influence expression of *shh* target and down-stream genes in the adult spinal cord, I performed RT-PCR to analyse expression of the *shh* target gene *patched1* (Sanchez and Ruiz i Altaba, 2005) and of *olig2*, expression of which depends on *shh* during development (Lu et al., 2000). Expression of both *patched1* and *olig2* mRNA was clearly reduced in the lesioned spinal cord after the cyclopamine treatment, compared to tomatidine treatment (Fig. 24). The PCR was normalized against the housekeeping gene *glyceraldehyde-3-phosphate dehydrogenase* (*GAPDH*). This suggests that cyclopamine affects ventricular proliferation and motor neuron differentiation by specifically blocking the *shh* pathway.

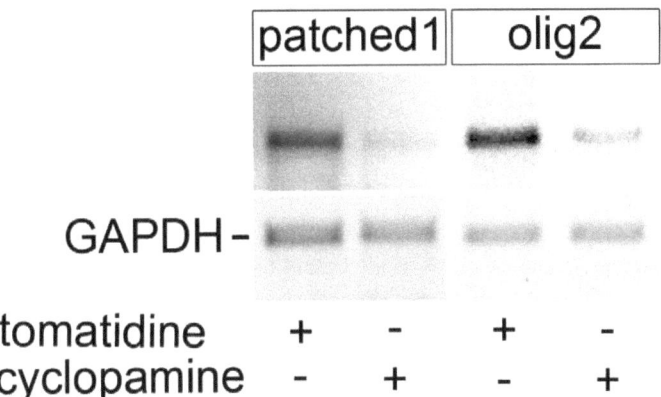

Fig 24: Intraperitoneal injection of cyclopamine reduces *patched1* and *olig2* expression in the lesioned spinal cord. A single injection of cyclopamine at 3 days post-lesion strongly reduces detectability of *patched1* and *olig2* by RT-PCR at 4 days post-lesion compared with tomatidine injection.

3.2 Motor axon pathfinding during development

To analyse the molecular mechanism of axonal differentiation of motor neurons, I turned to the axon growth of so-called primary motor neurons in embryonic zebrafish. The pioneering primary motor axons in the zebrafish trunk are guided by multiple cues along their pathways. We decided to analyse the function of plexins. Plexins are receptor components for semaphorins that influence motor axon growth and pathfinding. This study was a cooperation with Dr. Julia Feldner and is published (Feldner et al., 2007). I contributed mainly to cloning of *plexinA3* overexpression construct, localization of *plexinA3* mRNA in primary motor neurons during axon outgrowth and rescue experiments of morpholino phenotypes.

3.2.1 Cloning of *plexinA3*

A search of the Ensembl database (www.ensembl.org/Danio_rerio/) predicted ENSDARG00000016216 (Ensembl release 19) on zebrafish chromosome 8 to be most closely related to mouse and human *plexinA3* gene sequences. We then analysed the general domain structure of the deduced protein (1892 amino acids). It is identical to that of *plexinA3* in other vertebrate species: a Sema domain, followed by three MRS (Met Related Sequence) domains, four IPT (Immunoglobulin-like fold shared by Plexins and Transcription factors) motifs, and the characteristic intracellular Plexin domain at the C-terminus. The transmembrane domain of the zebrafish protein is located between the IPT motifs and the Plexin domain and comprises amino acids 1241-1263 (Fig. 26A). The protein sequence of the cloned gene has significant structural homology and overall amino acid identity (73%) with human (Maestrini et al., 1996) and mouse (Kameyama et al., 1996) *plexinA3*. Dr. J. Feldner showed in a phylogenetic tree, constructed using the Clustal method (Chenna et al., 2003), that zebrafish *plexinA3* segregated with *plexinA3* homologs of other species (Fig. 26, B). These data strongly suggest that we cloned a species homolog of *plexinA3*.

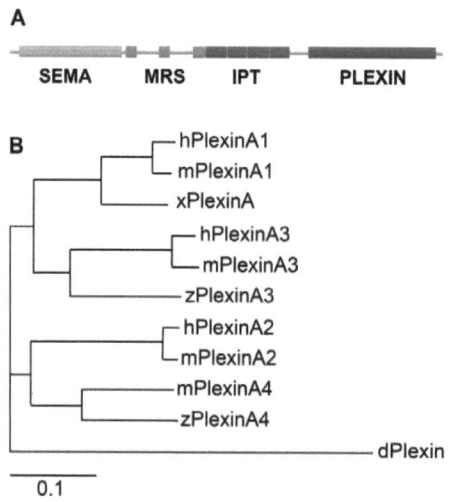

Fig. 26: Structual features and identitiy of *plexinA3* in zebrafish. **A:** Domain strucutre of *plexinA3*. SEMA, Sepmaphorin domain; PLEXIN, plexin domain; MRS, Met-related sequence. **B:** Multiple comparisons in a phylogenetic tree group zebrafish *plexinA3* with *plexinA3* homologs in other vertebrates. *Dosophila* was added as an outgroup.

The scalebar represents 10 substitutions per 100 aa. z, zebrafish; m, mouse; h, human; x, *Xenopus*; d, *Drosophila*. (Feldner et al., 2007).

3.2.2 *PlexinA3* is strongly expressed in spinal motor neurons

In situ hybridization indicated expression of *plexinA3* mRNA mainly in the developing nervous system (Fig. 27 A, B). A particularly strong signal was found in regular clusters of cells at the ventral edge of the spinal cord at 16 and 24 hpf, i.e. during the time of axon outgrowth of primary motor neurons. Double labeling of the mRNA with GFP immunohistochemistry in HB9:GFP transgenic animals at 24 hpf revealed co-localization of the mRNA in GFP$^+$ motor neuron clusters from which the CaP axon started to grow in developmentally younger caudal segments. Conspicuous *plexinA3* mRNA expression was also found in more dorsal GFP negative spinal neurons (Fig. 27 C-E). In more rostral segments in which the MiP axon could be seen to grow out, the mRNA was detectable in adjacent cells that probably represent the CaP and MiP primary motor neurons (Fig. 27 F-H). Cells in the extra-spinal pathway of motor axons did not express detectable levels of *plexinA3* mRNA. Thus, *plexinA3* mRNA is expressed in primary motor neurons during axon outgrowth.

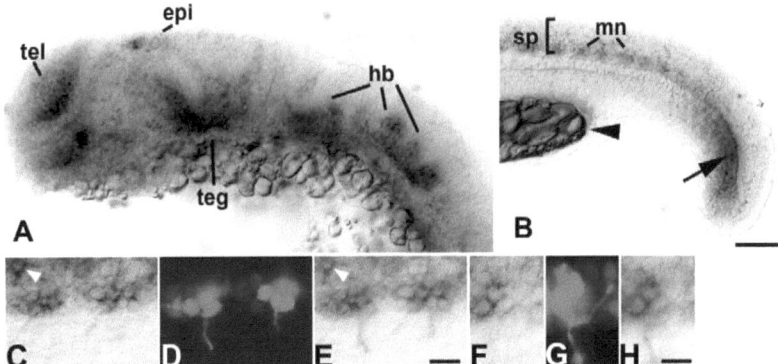

Fig 27: Expression pattern of *plexinA3* mRNA. Lateral views of whole mounted 24 hpf embryos are shown, rostral is left. **A,B:** *PlexinA3* mRNA is present in the telencephalon (tel), epiphysis (epi), tegmentum (teg), hindbrain neurons (hb) in the head (A), as well as in spinal cord (sp) and motor neurons (mn) (B). Additional expression is found in the tip of the tail (arrow in B). Yolk droplets (arrowhead in B) show non-specific staining. **C–H:** In situ hybridization of *plexinA3*-mRNA shows expression in clusters of GFP-

immunopositive motor neurons of *HB9:GFP* transgenic fish in lateral views at 24 hpf. C–E, A caudal region in which CaP axons (D, arrowheads) are just growing out. The arrowhead in C and E depicts a more dorsal, GFP-immuno negative cell that shows strong expression of *plexinA3*-mRNA. At higher magnification in **F–H**, two adjacent intensely *plexinA3* mRNA$^+$ neurons in a more rostral segment are depicted. These are likely the CaP (right cell) and MiP (left cell) motor neurons, judging by the trajectories of the GFP$^+$ MiP (arrow) and CaP (arrowhead) axons. Bars A,B = 100 µm, C-E = 25 µm, F-H = 12.5 µm.

3.2.3 *PlexinA3* is necessary for motor axon pathfinding

Ventral motor nerve growth in *plexinA3* morpholino injected embryos was analysed at 24 hpf using anti-tubulin immunohistochemistry (Fig. 28, A-F). Injection of 1 mM *plexinA3* morpholino1 led to abnormal growth of primary motor axons. Aberrations of ventral motor nerves, which have normally grown as one unbranched nerve beyond the ventral edge of the notochord at 24 hpf (Fig. 28 A,D), can be grouped into two categories: hemisegments showed an additional nerve exiting the spinal cord (Fig. 28 E, F) or nerves were abnormally branched (Fig. 28 B, C).

In 64% of the affected hemisegments, mostly one additional nerve of variable length grew ventrally from an additional exit point in the ventral spinal cord (Fig. 28 E, F). The additional nerve ran parallel to the main nerve or joined it at variable positions dorsal to the horizontal myoseptum. In 68% of the hemisegments showing additional exit points it could not be resolved whether the nerve emanated rostral or caudal to the segment border because the nerves grew very close to it. In the remaining hemisegments, 73% of the additional exit points were located in the posterior half of the somites, 25% were in the anterior half of the somites, or in both the anterior and posterior somite half (2%). On average, 4.7 ± 0.4 hemisegments/embryo had multiple exits in affected embryos.

Ventral motor nerves were aberrantly branched in 35% of the affected hemisegments (Fig. 28 C, D). The vast majority of these branches (82%) were directed caudally. Bifurcated (10%), rostrally (5%) and bilaterally (3%) branched nerves were observed less frequently. On average, 3.4 ± 0.2 hemisegments/embryo showed aberrant branching in affected embryos.

The effects were dose dependent with 26%, 43%, 64% of the embryos showing aberrant nerve branching and 18%, 56%, 94% of the embryos showing additional exit points from the spinal cord following injections of 0.25, 0.5, and 1 mM morpholino1, respectively. Injecting 1 mM morpholino2 phenocopied these effects (83% embryos affected by abnormal branching; 95% embryos affected by additional exits). Injections of 1 mM of a morpholino in which 5 bases were mismatched had no effect (14% embryos affected by branching, 12% embryos affected by additional exits).

Thus, knockdown of *plexinA3* induces both branching of ventral motor nerves and additional exit points from the spinal cord preferentially in the posterior half of the trunk segments.

To elucidate whether dorsal motor axons, which are obscured in anti-tubulin labelled embryos, were affected by the morpholino treatment we analysed *HB9:GFP* transgenic fish at 31 hpf. At this time point, GFP^+ axons had grown into the dorsal MiP pathway at the level of the yolk extension in uninjected animals (Fig. 28, G). In 1 mM *plexinA3* morpholino1 (n = 10 embryos) or morpholino2 (n = 13 embryos) injected *HB9:GFP* embryos, axons were also present in the MiP pathway, including the segments with multiple exits (n = 47 segments). Interestingly, in nine of these segments, the additional exit points of ventral motor axons also produced additional axons that grew dorsally (Fig. 28, H). Most of these dorsally growing axons were located more laterally than the normal MiP axons as determined in confocal image stacks (not shown). This indicates that these ectopic axons did not simply follow a MiP pathway. Branching away from the normal MiP pathway was also slightly increased by the morpholino treatment (Fig. 28 H). The frequency of dorsal motor nerves that were branched ventral to the level of GFP^+ ventral spinal neurons was 33.4% ± 2.84% hemisegments/embryo (n = 327 hemisegments) in morpholino treated animals and 12.1% ± 2.04% hemisegments/embryo (n = 215 hemisegments, Mann-Whitney U-test, $P < 0.0001$) in *HB9:GFP* embryos injected with 5 miss-match (mm) morpholino (n=14 embryos). Thus, additional nerves and increased nerve branching occur in both ventral and dorsal primary motor axon paths.

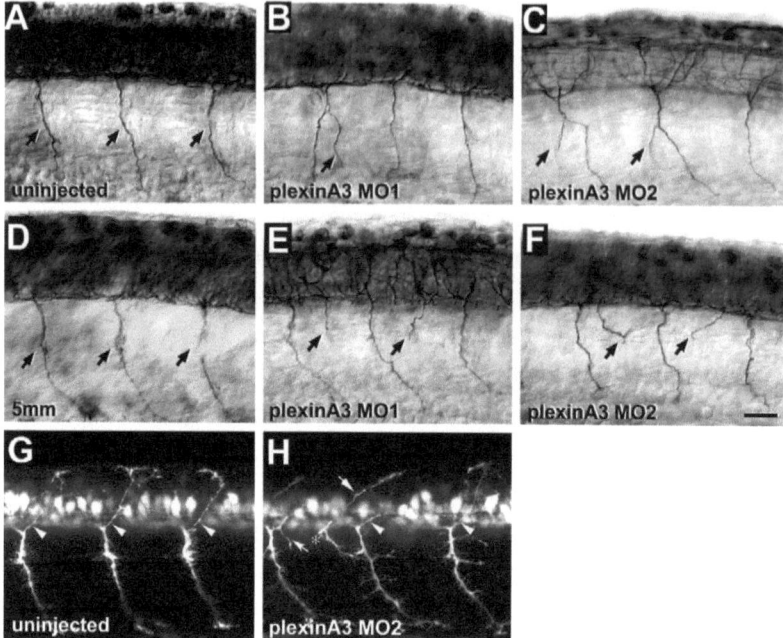

Fig. 28: Function of *plexinA3* in primary motor neurons. **A–F:** Lateral views at midtrunk levels of anti-tubulin-labelled whole mounted 24 hpf embryos are shown. In uninjected embryos (A) or those injected with 1 mM *plexinA3*, 5 mm morpholino (5 mm; D), single unbranched motor nerves (arrows in A and D) grow ventrally out of the spinal cord. Injection of 1mM *plexinA3* morpholino1 (MO1) induces branching (arrow in B) or a second spinal exit point for motor nerves per hemisegment (arrows indicate additional nerves in E). Injection of 1 mM *plexinA3* morpholino2 (MO2) also induced aberrant branching (arrows in C) of the ventral motor nerve and additional nerves exiting the spinal cord (arrows in F). **G,H:**, Axons in the dorsal MiP pathway are visualized in *HB9:GFP* transgenic fish in selected confocal image stacks at 31 hpf, indicating normal growth in uninjected embryos (arrowheads in G), and excessive branching (curved arrow in H) and supernumerary nerves (straight arrow in H) in 1 mM *plexinA3* morpholino2-injected embryos. The asterisk in H indicates an additional nerve exit point with a dorsal and ventral nerve branch exiting the spinal cord. Arrowheads in H point to normal appearance of axons in the dorsal motor axon pathway. Rostral is left in A to P. Bar 25µm.

3.2.4 *PlexinA3* morpholino phenotypes are specific.

Morpholino phenotypes need to be carefully controlled. One of the best controls is to rescue the morpholino-induced phenotype with a co-injected full length RNA of the targeted gene, which does not include a binding sequence for the morpholino. Overexpression of a full length myc-tagged *plexinA3* mRNA alone had no effect on motor axon growth as determined by anti-tubulin immunohistochemistry at 24 hpf (data not shown). However, co-injection of *plexinA3* morpholino2 (titrated to 0.3 mM), with *plexinA3* mRNA, which does not have a binding sequence for the morpholino, led to a strong and significant reduction in the frequency of both abnormal branching (13% affected embryos) and additional exits (16% affected embryos). This was compared to injection of 0.3 mM *plexinA3* morpholino2 alone at 24 hpf (embryos affected by branching: 87%, $P < 0.0001$; embryos affected by additional exits: 49%, $P < 0.01$; Tab. 3). This indicates that morpholino phenotypes are most likely due to reduction of *plexinA3* expression.

Tab. 3: Effects of morpholino treatment on ventral motor nerves:

Injection type PLEXINA3 MORPHOLINOS	n	Embryos with aberrant ventral motor nerve branching (%)	Embryos with additional exits of ventral motor nerves (%)
Vehicle	53	11.5 ± 7.3	4.4 ± 2.4
PlexinA3 5mm MO (1 mM)	51	13.8 ± 5.3	12.0 ± 0.3
PlexinA3 MO1 (0.25 mM)	53	26.0 ± 10.4	17.5 ± 5.2
PlexinA3 MO1 (0.5 mM)	65	43.2 ± 14.7 **	56.0 ± 11.7 ***
PlexinA3 MO1 (1 mM)	68	63.8 ± 7.3 ***	93.9 ± 2.7 ***
PlexinA3 MO2 (1 mM)	66	82.9 ± 6.5 ***	94.7 ± 2.5 ***
„RESCUE" EXPERIMENTS			
PlexinA3 MO2 (0.3 mM) alone	42	87.3 ± 3.5	48.8 ± 12.5
PlexinA3 MO2 (0.3 mM) + PlexinA3 mRNA	47	12.7 ± 4.4***	15.7 ± 15.7**

Morpholino doses are indicated in brackets. n = numbers of embryos analyzed, MO = morpholino, plexinA3 MO1/MO2: morpholino1/2 against plexinA3, plexinA3 5mm MO: morpholino with 5 mismatched bases based on plexinA3 morpholino1, ** = $P < 0.01$, *** = $P < 0.001$ (Fisher's Exact Test).

In situ hybridization already indicated that *plexinA3* is mainly expressed in the spinal motor neurons and not in cells in other axonal pathways. However to analyse whether changes in the axonal pathways might have induced aberrant axon growth, we analysed other spinal neurons and trunk structures. Analysis of markers of the horizontal and vertical myosepta, as well as spinal floorplate, motor neuron somata, commissural primary ascending interneurons in the spinal cord and Mauthner neurons with their spinal axons, indicated normal differentiation of these structures after injection of 1mM *plexinA3* morpholino1, including those hemisegments with aberrant motor axon growth (Feldner et al. 2007). Thus, the spinal environment of primary motor axons was not detectably altered by the morpholino injections. Hence, we identified *plexinA3* as a crucial receptor in motor neurons for correct pathfinding of primary motor axons in embryonic zebrafish.

4 DISCUSSION

4.1 Adult zebrafish are capable of motor neuron regeneration

In adult zebrafish, a spinal lesion triggers neural stem cells in the spinal cord to produce motor neurons. These cells are added to pre-existing spinal tissue adjacent to a spinal lesion site, as evidenced by the presence of myelin debris at these levels and increased ventricular proliferation in a region covering more than a third of the entire spinal cord around the actual lesion site. In the lesion site itself, normal cytoarchitecture appears not to be restored. Thus, this model differs significantly from tail regeneration paradigms in amphibians in which the entire spinal cord tissue is completely reconstructed from an advancing blastema (Echeverri and Tanaka, 2002).

Two proliferation studies, one with the permanent marker BrdU and the other with a marker for acute proliferation (*PCNA*), revealed the time course and the distribution of newborn cells in the lesioned spinal cord. Both proliferation markers showed a strong increase in the number of labelled cells after a lesion. While BrdU labelled cells were found at the ventricle and in the parenchyma, the increase of *PCNA* labelled cells was only significant in the ventricular zone. These findings suggest that ventricular cells give rise to the majority of newborn cells after a lesion. Newborn cells at the ventricle may subsequently migrate into the parenchyma. Highest proliferative activity was detected close to the lesion site, which is consitent with proliferation being a specific lesion-induced response of the zebrafish spinal cord.

I focused on motor neurons in order to limit the scope of the analysis of different cell types that could theoretically could be newly generated after a lesion. This was because motor neuron loss is the major problem in amyotrophic lateral sclerosis (ALS) and other motor neuron diseases (Ryu and Ferrante, 2007), as well as one of the first problems that needs to be solved after a spinal cord injury. Moreover, the sequence of motor neuron differentiation during development is well-established and markers for motor neuron differentiation are available, such as antibodies (against ChAT, HB9 and islet-1/-2) and two independent transgenic reporter lines in which motor neurons are labelled (HB9:GFP and islet-1:GFP) (Higashijima et al., 2000; Flanagan-Steet et al., 2005).

Looking at the transgenic HB9:GFP line we expected that the numbers of the GFP^+ motor neurons would be transiently reduced after a lesion and increase again to levels comparable to those in unlesioned animals. However, $HB9:GFP^+$ motor neurons had more complex reaction patterns than originally anticipated. In the unlesioned situation the majority of $HB9:GFP^+$ motor neurons were large (> 12 µm diameter) and mostly immunopositive for the mature motor neuron marker ChAT. Only very few smaller motor neurons (< 12 µm diameter) where present. The situation changed drastically after a lesion; the number of large motor neurons decreased significantly within the first week. In contrast small $HB9:GFP^+$ motor neurons gradually increased from 20.0 ± 7.66 in the unlesioned situation to a maximal value of 869.5 ± 106.78 at 2 wpl. These small motor neurons were rarely $ChAT^+$ or decorated by $SV2^+$ contacts and were reduced in number in the following weeks. At the same time the number of large $HB9:GFP^+$ motor neurons increased again. This result was confirmed with the transgenic islet-1:GFP line, in which motor neurons also express GFP. In this line, the number of small motor neurons likewise increased transiently after a lesion while large neurons disappeared. These observations led us to hypothesize that small HB9:GFP or islet-1:GFP positive neurons were immature motor neurons that were derived from the proliferating ventricular cells and in the process of regeneration matured to large differentiated motor neurons.

Indeed, intraperitoneal BrdU injections indicated that small motor neurons were newly generated after a lesion. The injections labelled 23% of the small, GFP^+ motor neurons with BrdU. Repeated BrdU injections kills the fish, therefore only three injections could be administered, one every other day. However, the bioavailability of BrdU has been estimated to be only 4 hours after injection (Zupanc and Horschke, 1995). This is one explanation why less than a quarter of the small motor neurons were double labelled. Another reason might be that after labelling with BrdU the number of cell divisions that occurred diluted the BrdU labelling below the detection level of immunohistochemistry.

Nevertheless, the numbers of small $HB9:GFP^+$ motor neurons increased from 20 in unlesioned animals to 870 at 2 wpl, which represents a 43-fold increase. Moreover, a large proportion of these cells were $BrdU^+$. This suggests that nearly all small motor neurons at 2 wpl were newly generated.

Islet-1 and *HB9* are markers for different subpopulations and/or differentiation stages of motor neurons in the developing spinal cord of amniotes (William et al., 2003). In this study I found evidence that this may also be the case for the regenerating spinal cord of adult zebrafish, by using combinations of the *islet-1:GFP* and *HB9:GFP* transgenic reporter lines and antibodies to *islet-1/-2* and *HB9*. In the lesioned situation, the *islet-1/-2* antibody recognizes the majority of all *islet-1:GFP* labelled small motor neurons and additionally a huge population of cells that do not express the *islet1* transgene. These are thought to be $islet-2^+$ neurons. Combining the *HB9:GFP* line with the *islet-1/-2* antibody revealed that there is a large populations of motor neurons that are only labelled with the *HB9:GFP* transgene. Similarly, the *HB9* antibody labelled mainly GFP^- cells in the *islet-1:GFP* line (data not shown). This suggests heterogeneity among the newly generated motor neurons in terms of their marker expression profile. These findings indicate that spinal cord lesion in zebrafish may induce the generation of diverse motor neuron cell types.

One important question is whether newborn motor neurons are integrated into the spinal cord circuitry and fulfill their purpose as motor neurons. According to a time course study of *ChAT* immunohistochemistry conducted by a postgraduate student in our laboratory, Veronika Kuscha, the number of fully differentiated, $ChAT^+$ cells drops from 478.0 ± 111.12 (n = 3 animals) cells in unlesioned animals to 234.7 ± 111.12 (n = 3 animals) at 2 wpl and recovers by 6 wpl to 348.4 ± 67.27 (n = 4 animals) in the 1500 μm around the lesion site. Hypothesizing that the increase in $ChAT^+$ cells is due to the maturation of the newborn motor neurons (e.g. small $HB9:GFP^+$ cells), 114 $ChAT^+$ cells must have been newly generated between 2 and 6 wpl. Taking into account that the BrdU injection scheme labels almost a quarter of all newborn motor neurons, we expect to find 28.5 $ChAT^+/BrdU^+$ double labelled cells within 1500 μm around the lesion site. Confirming this expectation, a triple-immunohistochemisty experiment using antibodies against *ChAT*, BrdU and the synaptic marker *SV2* reveals that at 6 wpl 29.3 ± 23.14 (n = 3 animals) cells are double labelled for *ChAT* and BrdU and are decorated by $SV2^+$ contacts within 1500 μm around the lesion site. This suggests that after a spinal cord lesion the loss of mature motor neurons is compensated for by maturation and integration of newly generated motor neurons. The presence of $SV2^+$ contacts on these neurons indicates that they may receive synaptic input and may thus be integrated into the spinal circuitry. Furthermore, double labelling with BrdU and retrograde tracing from the muscle tissue reveals that a newborn motor neuron contacted muscle tissue at 8 wpl, which is consistent with a motor neuron function of these newly generated cells. All these lesion induced morphological changes match the time course of functional recovery which plateaus at 6 wpl (Becker et al., 2004). Therefore, it is possible that newly generated fully mature motor neurons replace lost motor neurons and contribute to functional recovery after a spinal cord lesion.

4.2 *Olig2⁺* ependymo-radial glial cells are the putative stem cells in adult motor neuron regeneration

Some observations suggest that *olig2⁺* ependymo-radial glial cells are neural stem cells. Continued low level proliferation in the unlesioned adult (Park et al., 2007) and increased proliferation after a lesion do not lead to significant changes in the number of *olig2:GFP⁺* cells, suggesting asymmetric cell divisions and some potential for self-renewal. Using a BrdU label retention experiment, I was able to confirm that *olig2:GFP⁺* ependymo-radial glial cells were label-retaining, which is another stem cell characteristic (Grandel et al., 2006).

Moreover, these cells express the stem cell markers *BLBP* and *atypical protein kinase C* protein (Park et al., 2007). A stem cell role for olig⁺ ependymo-radial glia cells would be in agreement with that of several other radial glia cell types in developing mammals and in adult zebrafish (Pinto and Götz, 2007). For example, Müller cells, the radial glia cell type in the adult retina, can produce different cell types in adult zebrafish, depending on which of these are lost after specific lesions (Bernardos et al., 2007; Fimbel et al., 2007).

Furthermore, lineage tracing experiments showed *olig2:GFP*$^+$ ependymo-radial glial cells already expressed *HB9* and were BrdU$^+$ in a triple labelling experiment and were *islet-1/-2*$^+$ in double labelling experiments. This was only observed after lesion as *olig2:GFP*$^+$ ependymo-radial glial cells were never found to produce neurons in the unlesioned situation. Therefore, *olig2* expressing ependymo-radial glial cells switch from a gliogenic phenotype to motor neuron production, indicating that these are the likely adult stem cells. So far a fate switch for spinal progenitor cells has only been described for generating neurons to generating glial cells during development in different vertebrates (Götz and Barde, 2005).

4.3 Mechanisms of motor neuron regeneration in adult zebrafish are similar to developmental mechanisms

Comparing developmental mechanisms of motor neuron formation with motor neuron regeneration revealed similarities. In the development of vertebrate spinal cord motor neurons the morphogen *Sonic Hedgehog* (*shh*) sets a dorso-ventral gradient which leads to five progenitor domains. Each of these domains is defined by the expression of a unique set of transcription factors (Jessell, 2000). The *olig2*$^+$ domain in combination with the transcription factors *nkx6.1* and *pax 6* defines the pMN progenitor cell domain that gives rise to motor neurons.

The ventricular zone of the adult spinal cord in zebrafish retains its embryonic polarity. The expression domains of *shh*, *nkx6.1* and *olig2* around the ventral ventricle are comparable to those in the embryonic neural tube (Fuccillo et al., 2006). Even though expression of these markers was strongly increased after a lesion, they were expressed in a comparable pattern in unlesioned animals. The transcription factor *olig2* is detectable as a transgene but is below the detection threshold of in situ hybridization in unlesioned animals (Park et al., 2007).

The polarized increase of *shh* profoundly influences motor neuron regeneration. I tested this by using small molecule perturbation of *shh* function. We used the well established inhibitor cyclopamine, which also abrogates differentiation of motor neurons in developing zebrafish (Park et al., 2004; Fuccillo et al., 2006).

Injecting the substance into adult zebrafish specifically reduced expression of the *shh* target gene *patched1* and the down-stream gene *olig2* in semi-quantative PCR, indicating specific action of the compound on the *shh* pathway. Cyclopamine reduced ventricular proliferation and the number of regenerated motor neurons, suggesting that *shh* signaling is necessary for motor neuron regeneration.

So far, marker expression and *shh* signaling is remarkably similar between regeneration and developmental processes. However, the fate switch from gliogenesis to motor neuron production in progenitor cells appears to be specific for regeneration (Raya et al., 2003).

4.4 Implications of motor neuron regeneration in zebrafish for spinal cord regeneration in mammals

We were able to investigate proliferation of endogenous progenitor cells, neuronal differentiation signals and motor neuron regeneration in the injured spinal cord of the adult zebrafish, which shows successful recovery of function. In mammals, spinal cord injury leads to extensive secondary cell loss around the lesion site (Demjen et al., 2004). In the lesioned spinal cord of mammals, proliferation and expression of nestin, an intermediate filament marker for progenitor cells, is increased around the ventricle and in parenchymal astrocytes, some of which carry radial processes (Yamamoto et al., 2001; Shibuya et al., 2002). Exogenous administration of *shh* to the lesioned spinal cord increases precursor cell proliferation (Bambakidis and Miller, 2004). Expression of *pax6*, a transcription factor of progenitor cells in the embryonic spinal cord (Fuccillo et al., 2006), is increased in the ependymal layer of the lesioned adult mammalian spinal cord. However, contrary to development, this expression is not polarized. *Olig2* and several other factors are not re-expressed (Yamamoto et al., 2001) and no neurons were formed. One report (Chen et al., 2005) showed a surprisingly wide spread increase in *shh* mRNA expression in the entire ependyma and several parenchymal cell types in mice. Overall, these observations suggest that spinal progenitors in mammals show some plasticity after a lesion and could be induced to produce new motor neurons. This might be achieved by combining *shh* with other growth factors (Ohori et al., 2006).

It will be very interesting to elucidate whether other signaling pathways are involved in neuronal regeneration in the adult spinal cord (i.e. notch pathway, retinoic acid pathway and wnt pathway), either directly or indirectly by stimulating *shh* expression. Identifying signals that trigger and control neuronal replacement from endogenous progenitor cells in fish may inform future cell therapies for spinal cord injury, but also for neurodegenerative diseases, such as ALS (Roskams and Tetzlaff, 2005).

4.5 *PlexinA3* is crucial for motor axon pathfinding

To understand motor neuron regeneration in the adult spinal cord it is important also to analyse motor neuron differentiation during development. We analysed the role of *plexinA3* for axonal differentiation of motor neurons. Morpholino knockdown experiments suggest that *plexinA3* in dorsal and ventral motor axons may be necessary to correctly read repellent cues from semaphorins during axon outgrowth. The observed phenotypes are consistent with those observed for knockdown of NRP1a in our lab (Bovenkamp et al., 2004), indicating that NRP/plexin receptor complexes are likely to exist in primary motor axons. The receptor knockdown phenotypes observed, additional exits from the spinal cord and branching of the ventral and dorsal motor nerve, are consistent with a release of axon growth from environmental restrictions. Indeed, class 3 semaphorins are expressed in the trunk environment and are thought to signal through plexin receptors (Birely et al., 2005; Gulati-Leekha and Goldman, 2006).

Up to 95% of *plexinA3* morpholino-injected embryos show specific types of motor axon aberrations and 30% of all hemi-segments analysed were aberrant. This is more than in comparable studies of other proteins in motor axon growth (Feldner et al., 2005; Sato-Maeda et al., 2006). Two sequence-independent morpholinos yielded identical results and I was able to rescue all of the phenotypes to a significnat degree by supplementing *plexinA3* using mRNA overexpression. Using various markers, we could not find detectable changes in the spinal cord and trunk structures of morpholino treated embryos. This suggests a major and specific function of *plexinA3* in primary motor neurons.

We conclude that growth and pathfinding of primary motor axons in zebrafish is governed by a complex interplay of different semaphorin ligands and receptors of which *plexinA3* is a crucial component. Shortly after completion of our *plexinA3* knockdown study, two independent publications reported *plexinA3* mutant zebrafish with almost identical phenotypes (Palaisa and Granato, 2007; Tanaka et al., 2007). This shows that forward and reverse genetics approaches have the potential to elucidate motor axon differentiation in this in vivo model in unprecedented detail. It will be interesting to determine to what degree adult regenerating motor neurons recapitulate embryonic gene expression.

4.6 Conclusion

We conclude that the zebrafish, a powerful genetic model which is accessible to pharmacological manipulations, provides an opportunity to identify the evolutionarily conserved signals that control embryonic motor neuron differentiation and massive regeneration of motor from endogenous stem cells in the adult spinal cord.

5 SUMMARY

Zebrafish, in contrast to mammals, are capable of functional spinal cord regeneration. Spinal motor neurons are major targets for axons regenerating from the brainstem. Using immunohistochemical markers and transgenic reporter fish for motor neuron markers (*HB9*, *islet-1*), this study demonstrates that large differentiated motor neurons are transiently lost after a spinal lesion, suggesting that these cells undergo cell death after a lesion and may be replaced by proliferation. Indeed, a massive and transient increase in the number of small, undifferentiated motor neurons, which were labelled by the proliferation marker bromodeoxyuridine, was observed. Proliferation and lineage tracing studies indicated significant proliferation only at the spinal ventricle and that a subset of *olig2* expressing ependymo-radial glial cells are the likely motor neuron progenitor/stem cells in the lesioned spinal cord.

A spinal lesion increased expression of *sonic hedgehog* (*shh*), an embryonic differentiation signal for motor neurons. Blocking this signal with an antagonist reduced progenitor cell proliferation and motor neuron differentiation. This suggests that *shh* is an important signal for motor neuron differentiation during adult motor neuron regeneration.

To learn more about axonal differentiation of motor neurons, the role of the cell recognition molecule *plexinA3* was investigated during the outgrowth of embryonic primary motor axons. The molecule is selectively expressed in primary motor neurons. Knockdown of expression led to ectopic exiting from the spinal cord and excessive branching of motor axons. Over-expression of full length *plexinA3* rescued this effect, indicating specificity of experimental manipulations. Thus, *plexinA3* expression is crucial for motor axon pathfinding during development.

Overall, this study demonstrates that adult zebrafish are capable of motor neuron regeneration from endogenous progenitor/stem cells and that *shh* is an important regulator of motor neuron regeneration. *PlexinA3* is crucial for motor axon differentiation in embryonic zebrafish. This study establishes adult spinal cord lesion as a model system for motor neuron regeneration, which may ultimately help to find ways to promote motor neuron regeneration also in human conditions, such as spinal cord injury or motor neuron disease.

6 LITERATURE

Bambakidis NC, Miller RH (2004) Transplantation of oligodendrocyte precursors and sonic hedgehog results in improved function and white matter sparing in the spinal cords of adult rats after contusion. Spine J 4:16-26.

Bambakidis NC, Butler J, Horn EM, Wang X, Preul MC, Theodore N, Spetzler RF, Sonntag VK (2008) Stem cell biology and its therapeutic applications in the setting of spinal cord injury. Neurosurg Focus 24:E20.

Beattie CE (2000) Control of motor axon guidance in the zebrafish embryo. Brain Res Bull 53:489-500.

Beck CW, Christen B, Slack JM (2003) Molecular pathways needed for regeneration of spinal cord and muscle in a vertebrate. Dev Cell 5:429-439.

Becker CG, Becker T, eds (2007) Model Organisms in Spinal Cord Regeneration.

Becker CG, Lieberoth BC, Morellini F, Feldner J, Becker T, Schachner M (2004) L1.1 is involved in spinal cord regeneration in adult zebrafish. J Neurosci 24:7837-7842.

Becker T, Becker CG (2001) Regenerating descending axons preferentially reroute to the gray matter in the presence of a general macrophage/microglial reaction caudal to a spinal transection in adult zebrafish. J Comp Neurol 433:131-147.

Becker T, Lieberoth BC, Becker CG, Schachner M (2005) Differences in the regenerative response of neuronal cell populations and indications for plasticity in intraspinal neurons after spinal cord transection in adult zebrafish. Mol Cell Neurosci 30:265-278.

Becker T, Wulliman MF, Becker CG, Bernhardt RR, Schachner M (1997) Axonal regrowth after spinal cord transection in adult zebrafish. J Comp Neurol 377:577-595.

Bernardos RL, Barthel LK, Meyers JR, Raymond PA (2007) Late-stage neuronal progenitors in the retina are radial Muller glia that function as retinal stem cells. J Neurosci 27:7028-7040.

Bernhardt RR (1999) Cellular and molecular bases of axonal regeneration in the fish central nervous system. Exp Neurol 157:223-240.

Birely J, Schneider VA, Santana E, Dosch R, Wagner DS, Mullins MC, Granato M (2005) Genetic screens for genes controlling motor nerve-muscle development and interactions. Dev Biol 280:162-176.

Bovenkamp DE, Goishi K, Bahary N, Davidson AJ, Zhou Y, Becker T, Becker CG, Zon LI, Klagsbrun M (2004) Expression and mapping of duplicate neuropilin-1 and neuropilin-2 genes in developing zebrafish. Gene Expr Patterns 4:361-370.

Briscoe J, Ericson J (2001) Specification of neuronal fates in the ventral neural tube. Curr Opin Neurobiol 11:43-49.

Carulli D, Laabs T, Geller HM, Fawcett JW (2005) Chondroitin sulfate proteoglycans in neural development and regeneration. Curr Opin Neurobiol 15:116-120.

Chapouton P, Adolf B, Leucht C, Tannhauser B, Ryu S, Driever W, Bally-Cuif L (2006) her5 expression reveals a pool of neural stem cells in the adult zebrafish midbrain. Development 133:4293-4303.

Chen J, Leong SY, Schachner M (2005) Differential expression of cell fate determinants in neurons and glial cells of adult mouse spinal cord after compression injury. Eur J Neurosci 22:1895-1906.

Chenna R, Sugawara H, Koike T, Lopez R, Gibson TJ, Higgins DG, Thompson JD (2003) Multiple sequence alignment with the Clustal series of programs. Nucleic Acids Res 31:3497-3500.

Coggeshall RE, Lekan HA (1996) Methods for determining numbers of cells and synapses: a case for more uniform standards of review. J Comp Neurol 364:6-15.

Demjen D, Klussmann S, Kleber S, Zuliani C, Stieltjes B, Metzger C, Hirt UA, Walczak H, Falk W, Essig M, Edler L, Krammer PH, Martin-Villalba A (2004) Neutralization of CD95 ligand promotes regeneration and functional recovery after spinal cord injury. Nat Med 10:389-395.

Dijkers MP (2005) Quality of life of individuals with spinal cord injury: a review of conceptualization, measurement, and research findings. J Rehabil Res Dev 42:87-110.

Echeverri K, Tanaka EM (2002) Ectoderm to mesoderm lineage switching during axolotl tail regeneration. Science 298:1993-1996.

Ecke I, Rosenberger A, Obenauer S, Dullin C, Aberger F, Kimmina S, Schweyer S, Hahn H (2008) Cyclopamine treatment of full-blown Hh/Ptch-associated RMS partially inhibits Hh/Ptch signaling, but not tumor growth. Mol Carcinog 47:361-372.

Eisen JS, Myers PZ, Westerfield M (1986) Pathway selection by growth cones of identified motoneurones in live zebra fish embryos. Nature 320:269-271.

Eisen JS, Pike SH, Romancier B (1990) An identified motoneuron with variable fates in embryonic zebrafish. J Neurosci 10:34-43.

Feldner J, Reimer MM, Schweitzer J, Wendik B, Meyer D, Becker T, Becker CG (2007) PlexinA3 restricts spinal exit points and branching of trunk motor nerves in embryonic zebrafish. J Neurosci 27:4978-4983.

Feldner J, Becker T, Goishi K, Schweitzer J, Lee P, Schachner M, Klagsbrun M, Becker CG (2005) Neuropilin-1a is involved in trunk motor axon outgrowth in embryonic zebrafish. Dev Dyn 234:535-549.

Fimbel SM, Montgomery JE, Burket CT, Hyde DR (2007) Regeneration of inner retinal neurons after intravitreal injection of ouabain in zebrafish. J Neurosci 27:1712-1724.

Flanagan-Steet H, Fox MA, Meyer D, Sanes JR (2005) Neuromuscular synapses can form in vivo by incorporation of initially aneural postsynaptic specializations. Development 132:4471-4481.

Fuccillo M, Joyner AL, Fishell G (2006) Morphogen to mitogen: the multiple roles of hedgehog signalling in vertebrate neural development. Nat Rev Neurosci 7:772-783.

Giger RJ, Cloutier JF, Sahay A, Prinjha RK, Levengood DV, Moore SE, Pickering S, Simmons D, Rastan S, Walsh FS, Kolodkin AL, Ginty DD, Geppert M (2000) Neuropilin-2 is required in vivo for selective axon guidance responses to secreted semaphorins. Neuron 25:29-41.

Götz M, Barde YA (2005) Radial glial cells defined and major intermediates between embryonic stem cells and CNS neurons. Neuron 46:369-372.

Grandel H, Kaslin J, Ganz J, Wenzel I, Brand M (2006) Neural stem cells and neurogenesis in the adult zebrafish brain: origin, proliferation dynamics, migration and cell fate. Dev Biol 295:263-277.

Gulati-Leekha A, Goldman D (2006) A reporter-assisted mutagenesis screen using alpha 1-tubulin-GFP transgenic zebrafish uncovers missteps during neuronal development and axonogenesis. Dev Biol 296:29-47.

Halloran MC, Sato-Maeda M, Warren JT, Su F, Lele Z, Krone PH, Kuwada JY, Shoji W (2000) Laser-induced gene expression in specific cells of transgenic zebrafish. Development 127:1953-1960.

Hewitson TD, Bisucci T, Darby IA (2006) Histochemical localization of apoptosis with in situ labeling of fragmented DNA. Methods Mol Biol 326:227-234.

Higashijima S, Hotta Y, Okamoto H (2000) Visualization of cranial motor neurons in live transgenic zebrafish expressing green fluorescent protein under the control of the islet-1 promoter/enhancer. J Neurosci 20:206-218.

Horner PJ, Power AE, Kempermann G, Kuhn HG, Palmer TD, Winkler J, Thal LJ, Gage FH (2000) Proliferation and differentiation of progenitor cells throughout the intact adult rat spinal cord. J Neurosci 20:2218-2228.

Huber AB, Kania A, Tran TS, Gu C, De Marco Garcia N, Lieberam I, Johnson D, Jessell TM, Ginty DD, Kolodkin AL (2005) Distinct roles for secreted semaphorin signaling in spinal motor axon guidance. Neuron 48:949-964.

J Sambrook, E F Fritsch, Maniatis T (1989) Molecular Cloning. In: New York Cold Spring Harbour Laboratory Press.

Jessell TM (2000) Neuronal specification in the spinal cord: inductive signals and transcriptional codes. Nat Rev Genet 1:20-29.

Johansson BB (2007) Regeneration and plasticity in the brain and spinal cord. J Cereb Blood Flow Metab 27:1417-1430.

Kameyama T, Murakami Y, Suto F, Kawakami A, Takagi S, Hirata T, Fujisawa H (1996) Identification of plexin family molecules in mice. Biochem Biophys Res Commun 226:396-402.

Kibbelaar RE, Moolenaar CE, Michalides RJ, Bitter-Suermann D, Addis BJ, Mooi WJ (1989) Expression of the embryonal neural cell adhesion molecule N-CAM in lung carcinoma. Diagnostic usefulness of monoclonal antibody 735 for the distinction between small cell lung cancer and non-small cell lung cancer. J Pathol 159:23-28.

Kirsche W (1950) Die regenerativen Vorgaenge am Rueckenmark erwachsener Teleostier nach operativer Kontinuitaetstrennung. In, pp 190-260. Berlin.

Kruger RP, Aurandt J, Guan KL (2005) Semaphorins command cells to move. Nat Rev Mol Cell Biol 6:789-800.

Kullander K (2007) Genetic Approaches to Spinal Locomotor Function in Mammals. In: Model Organisms in Spinal Cord Regeneration.

Lewis KE, Eisen JS (2001) Hedgehog signaling is required for primary motoneuron induction in zebrafish. Development 128:3485-3495.

Liu R, Cai J, Hu X, Tan M, Qi Y, German M, Rubenstein J, Sander M, Qiu M (2003) Region-specific and stage-dependent regulation of Olig gene expression and oligodendrogenesis by Nkx6.1 homeodomain transcription factor. Development 130:6221-6231.

Lu QR, Yuk D, Alberta JA, Zhu Z, Pawlitzky I, Chan J, McMahon AP, Stiles CD, Rowitch DH (2000) Sonic hedgehog--regulated oligodendrocyte lineage genes encoding bHLH proteins in the mammalian central nervous system. Neuron 25:317-329.

Maestrini E, Tamagnone L, Longati P, Cremona O, Gulisano M, Bione S, Tamanini F, Neel BG, Toniolo D, Comoglio PM (1996) A family of transmembrane proteins with homology to the MET-hepatocyte growth factor receptor. Proc Natl Acad Sci U S A 93:674-678.

Malicki J, Jo H, Wei X, Hsiung M, Pujic Z (2002) Analysis of gene function in the zebrafish retina. Methods 28:427-438.

Miyashita T, Yeo SY, Hirate Y, Segawa H, Wada H, Little MH, Yamada T, Takahashi N, Okamoto H (2004) PlexinA4 is necessary as a downstream target of Islet2 to mediate Slit signaling for promotion of sensory axon branching. Development 131:3705-3715.

Myers PZ, Eisen JS, Westerfield M (1986) Development and axonal outgrowth of identified motoneurons in the zebrafish. J Neurosci 6:2278-2289.

Nasevicius A, Ekker SC (2000) Effective targeted gene 'knockdown' in zebrafish. Nat Genet 26:216-220.

Ninkovic J, Götz M (2007) Signaling in adult neurogenesis: from stem cell niche to neuronal networks. Curr Opin Neurobiol 17:338-344.

Nusslein-Volhard C (2002) Zebrafish (Practical Approach). In: Oxford University Press.

Ohori Y, Yamamoto S, Nagao M, Sugimori M, Yamamoto N, Nakamura K, Nakafuku M (2006) Growth factor treatment and genetic manipulation stimulate neurogenesis and oligodendrogenesis by endogenous neural progenitors in the injured adult spinal cord. J Neurosci 26:11948-11960.

Palaisa KA, Granato M (2007) Analysis of zebrafish sidetracked mutants reveals a novel role for Plexin A3 in intraspinal motor axon guidance. Development 134:3251-3257.

Park HC, Shin J, Appel B (2004) Spatial and temporal regulation of ventral spinal cord precursor specification by Hedgehog signaling. Development 131:5959-5969.

Park HC, Mehta A, Richardson JS, Appel B (2002) olig2 is required for zebrafish primary motor neuron and oligodendrocyte development. Dev Biol 248:356-368.

Park HC, Shin J, Roberts RK, Appel B (2007) An olig2 reporter gene marks oligodendrocyte precursors in the postembryonic spinal cord of zebrafish. Dev Dyn 236:3402-3407.

Pinto L, Götz M (2007) Radial glial cell heterogeneity--the source of diverse progeny in the CNS. Prog Neurobiol 83:2-23.

Poss KD, Wilson LG, Keating MT (2002) Heart regeneration in zebrafish. Science 298:2188-2190.

Raya A, Koth CM, Buscher D, Kawakami Y, Itoh T, Raya RM, Sternik G, Tsai HJ, Rodriguez-Esteban C, Izpisua-Belmonte JC (2003) Activation of Notch signaling pathway precedes heart regeneration in zebrafish. Proc Natl Acad Sci U S A 100 Suppl 1:11889-11895.

Renoncourt Y, Carroll P, Filippi P, Arce V, Alonso S (1998) Neurons derived in vitro from ES cells express homeoproteins characteristic of motoneurons and interneurons. Mech Dev 79:185-197.

Roos M, Schachner M, Bernhardt RR (1999) Zebrafish semaphorin Z1b inhibits growing motor axons in vivo. Mech Dev 87:103-117.

Roskams AJ, Tetzlaff W (2005) Directing stem cells and progenitor cells on the stage of spinal cord injury. Exp Neurol 193:267-272.

Rupp RA, Snider L, Weintraub H (1994) Xenopus embryos regulate the nuclear localization of XMyoD. Genes Dev 8:1311-1323.

Rutishauser U (2008) Polysialic acid in the plasticity of the developing and adult vertebrate nervous system. Nat Rev Neurosci 9:26-35.

Ryu H, Ferrante RJ (2007) Translational therapeutic strategies in amyotrophic lateral sclerosis. Mini Rev Med Chem 7:141-150.

Saiki RK, Scharf S, Faloona F, Mullis KB, Horn GT, Erlich HA, Arnheim N (1985) Enzymatic amplification of beta-globin genomic sequences and restriction site analysis for diagnosis of sickle cell anemia. Science 230:1350-1354.

Sanchez P, Ruiz i Altaba A (2005) In vivo inhibition of endogenous brain tumors through systemic interference of Hedgehog signaling in mice. Mech Dev 122:223-230.

Sato-Maeda M, Obinata M, Shoji W (2008) Position fine-tuning of caudal primary motoneurons in the zebrafish spinal cord. Development 135:323-332.

Sato-Maeda M, Tawarayama H, Obinata M, Kuwada JY, Shoji W (2006) Sema3a1 guides spinal motor axons in a cell- and stage-specific manner in zebrafish. Development 133:937-947.

Schwab ME (2004) Nogo and axon regeneration. Curr Opin Neurobiol 14:118-124.

Shibuya S, Miyamoto O, Auer RN, Itano T, Mori S, Norimatsu H (2002) Embryonic intermediate filament, nestin, expression following traumatic spinal cord injury in adult rats. Neuroscience 114:905-916.

Shirasaki R, Pfaff SL (2002) Transcriptional codes and the control of neuronal identity. Annu Rev Neurosci 25:251-281.

Spencer T, Domeniconi M, Cao Z, Filbin MT (2003) New roles for old proteins in adult CNS axonal regeneration. Curr Opin Neurobiol 13:133-139.

Tallafuss A, Bally-Cuif L (2003) Tracing of her5 progeny in zebrafish transgenics reveals the dynamics of midbrain-hindbrain neurogenesis and maintenance. Development 130:4307-4323.

Tanabe Y, William C, Jessell TM (1998) Specification of motor neuron identity by the MNR2 homeodomain protein. Cell 95:67-80.

Tanaka H, Maeda R, Shoji W, Wada H, Masai I, Shiraki T, Kobayashi M, Nakayama R, Okamoto H (2007) Novel mutations affecting axon guidance in zebrafish and a role for plexin signalling in the guidance of trigeminal and facial nerve axons. Development 134:3259-3269.

Taylor JS, Van de Peer Y, Meyer A (2001) Revisiting recent challenges to the ancient fish-specific genome duplication hypothesis. Curr Biol 11:R1005-1008.

Tsuchida T, Ensini M, Morton SB, Baldassare M, Edlund T, Jessell TM, Pfaff SL (1994) Topographic organization of embryonic motor neurons defined by expression of LIM homeobox genes. Cell 79:957-970.

van Raamsdonk W, Smit-Onel MJ, Diegenbach PC (1993) Metabolic profiles of white and red-intermediate spinal motoneurons in the zebrafish. Acta Histochem 95:129-138.

van Raamsdonk W, Maslam S, de Jong DH, Smit-Onel MJ, Velzing E (1998) Long term effects of spinal cord transection in zebrafish: swimming performances, and metabolic properties of the neuromuscular system. Acta Histochem 100:117-131.

Westerfield M (1989) The zebrafish book, a guide for the laboratory use of zebrafish (*Brachydanio rerio*). In: University of Oregon Press.

Westerfield M, McMurray JV, Eisen JS (1986) Identified motoneurons and their innervation of axial muscles in the zebrafish. J Neurosci 6:2267-2277.

William CM, Tanabe Y, Jessell TM (2003) Regulation of motor neuron subtype identity by repressor activity of Mnx class homeodomain proteins. Development 130:1523-1536.

Yamamoto S, Nagao M, Sugimori M, Kosako H, Nakatomi H, Yamamoto N, Takebayashi H, Nabeshima Y, Kitamura T, Weinmaster G, Nakamura K, Nakafuku M (2001) Transcription factor expression and Notch-dependent regulation of neural progenitors in the adult rat spinal cord. J Neurosci 21:9814-9823.

Yaron A, Huang PH, Cheng HJ, Tessier-Lavigne M (2005) Differential requirement for Plexin-A3 and -A4 in mediating responses of sensory and sympathetic neurons to distinct class 3 Semaphorins. Neuron 45:513-523.

Zupanc GK, Horschke I (1995) Proliferation zones in the brain of adult gymnotiform fish: a quantitative mapping study. J Comp Neurol 353:213-233.

7 APPENDIX

7.1 Abbreviations

aa	amino acid
A	adenine
Amp	ampicillin
ATP	adensoin triphosphate
bp	base pairs
BSA	bovine serum albumine
C	cytosine
°C	degree celsius
cDNA	complementary deoxyribonucleic acid
CTA	cytosine triphosphate
dATP	2`-desoxyadenosine triphosphate
dpf	days post fertilisation
dCTP	2`-desoxycytosine triphosphate
dGTP	2'-desoxyguanosine triphosphate
DMSO	dimethylsulfoxide
DNA	deoxyribonucleic acid
DNase	desoxyribonuclease
dNTP	2'-desoxyribonucleotide-5'-triphosphate
dpf	days post fertilization
dpl	days post-lesion
DTT	dithiothreitol
E. coli	Escherichia coli
Fig.	figure
EDTA	ethylendiamintetraacetic acid
g	gram
G	guanosine

h	hour
h	human
hpf	hour post fertilisation
kan	Kanamycin
kb	kilo base pairs
l	litre
LB medium	Luria Bertani medium
m	milli (10^{-3})
min	minute
mRNA	messenger ribonucleic acid
n	number of animals, nano (10^{-9})
PBS	phosphate buffer saline
PCR	polymerase chain reaction
RNA	ribonucleic acid
RNase	ribonuclease
RT	room temperature
s	second
T	thymine
Tab.	table
T_m	melting temperature
v/v	volume per volume
w/v	weight per volume
wpl	weeks post lesion
x g	g-force
µ	micro (10^{-6})

7.2 Morpholinos

PlexinA3 MO1:
5'-ATACCAGCAGCCACAAGGACCTCAT-3'
PlexinA3 MO2:
5'-AGCTCTTCCCTCAAGCGTATTCCAG-3'

PlexinA3 5mm MO:
5'-ATACCACCACCCAGAACGACCTGAT-3'

7.3 Overexpression-construct *plexinA3*

7.3.1 Primers used to clone *plexinA3* overexpression construct:
plexin A3 (BamHI) forward
5`- GTGGATCCATGAGGTCCTTGTGGCTG -3`
plexinA3 (BamHI) reverse
5`- TAGGATCCGCTGCTGCCAGACATCAG-3`

7.3.2 Sequence of the overexpression construct for *plexinA3*:
5'-GTGGATCCATGAGGTCCTTGTGGCTGCTGGTATTTTCCTTCTCTGTTTTG
ACTGGGACCAACATGGCATTTCCAATGATTCTGTCGGAGCGCCCTGAAGT
CACCGGGAGCTTCAAGGTTAAAGACACGAGTCTCACTCACCTCACAGTGC
ACCGCAAAACTGGTGAGGTGTTCGTGGGTGCTATAAACCGAGTCTACAAG
CTTTCTGCCAATCTCACCGAAACGCGTTCTCACCAGACCGGTCCCGTGGA
AGACAACGCCAAGTGCTATCCACCCCCCAGTGTACGAGCTTGCACGCAGA
AACTGGAGTCTACAGACAACGTCAACAAATTGCTGCTGGTTGATTATGCGG
GCAACCGTCTGGCGGCCTGTGGAAGCATCTGGCAGGGCGTGTGCCAGTT
CCTGCGGTTGGAAGATCTGTTCAAGCTTGGTGAACCACATCACCGTAAAG
AGCACTACCTCTCGGGAGCCAAAGAGTCTGATGGGATGGCTGGAGTCGTG
GTGGGTGATGATGACGGAGACTTGAAGAAGAAAAAGAAAGGTGGCAGTCG
ACTCTTCATTGGTGCTGCAATCGATGGCAAATCAGAGTATTTTCCAACCCT
CTCTAGCCGTAAACTGGTGGCGGATGAGGAAAGTGTTAACATGTTCAGTTT
GGTCTACCAAGATGAGTTTGTGTCTTCTCAAATCAAGATACCTTCAGACAC
CCTCTCTCAGTATCCCGCATTTGATATCTACTACGTCTACGGGTTCTCCAG
CCGGACTTACATCTATTTCCTCACTTTGCAACTGGATACTCAGCTCACTCA
GGTGGATGTGACGGGGGAGAAGTTCTTCACCTCAAAAATAGTCCGCATGT
GCTCCAATGACACTGAGTTTTACTCCTACGTAGAGTTCCCGCTTGGGTGCA
CCAAGGATGGCGTGGAATACAGACTTGTTCAAGCTGCCTACAAGCATCGT
CCTGGAAAGATTCTGGCACAGGCTTTGGGCCTGTCTGAGGATGAGGATGT

```
CCTGTTCGTGATCTTCTCCCAGGGTCAGAAGAACAGGGCTAACCCACCGA
GAGAAACAGTGCTGTGCCTCTTCACACTGCACCAGATTAACCTGGCCATG
CGAGAGAGGATCAAGTCATGCTACCGCGGAGAGGGAAAGCTGTCTCTGC
CGTGGTTGCTCAACAAGGAGCTGCCTTGCATTAATACGCCCAAGCAGATT
GGTGATGATTTCTGCGGCCTGGTCTTGAATCAGCCCCTTGGGGGATTGAT
GGTGATCGAGGGCATTCCTCTGTTTGACGACCGCACTGACGGCATGGCAT
CAGTGGCTGCATACACATACGGAGACCATTCGGTGGTGTTTGTGGGCACT
CGCAGCGGCCACCTCAAGAAGATTCGAGTGAATGGTGTTCCTCCGCCGTC
AGAAAACGCTTTGCTGTACGAGACCGTGACCGTTGTGGAGGGAAGCCCCA
TCCTGAGGGACATGGTGTTCAGTCCAGACTATCAGTACATCTATCTGCTGA
GCGACAAACAGGTGAGTCGTCTGCCGGTGGAGAGCTGTTCTCAGTACAGC
AGCTGTAAGACGTGTCTGGGCTCTGGAGATCCTCACTGCGGCTGGTGTGT
CCTGCATAACAAGTGCTCCAGAAAGGAGGCCTGTGAGAAGTGGGCCGAG
CCGCTTCACTTCAGTACAGAGCTGAAGCAGTGTGTGGACATTACCGTCACT
CCGGATAACATGTCTGTGACCTCCGTGTCTACACAGCTGAGTGTGAAGGT
GGCGAACGTCCCGAACCTCTCTGCGGGGGTGACGTGTGTGTTTGAGGAG
CTCACCGAGAGTCCAGGAGAAGTGCTGGCTGAAGGACAAATCCTCTGCAT
GTCCCCTTCCCTTCGGGACGTCCCGTCTGTCACTCAGGGATATGGCGATA
AACGGGTCGTGAAGCTTTCTCTGAAGTCCAAAGAGACGGGGCTCAAATTC
ATCACCACCGACTTCGTCTTCTACAACTGCAGCGTTCTGCAATCGTGTTCA
TCGTGTGTTAGCAGTTCTTTCCCTTGCAACTGGTGTAAATATCGCCACATC
TGCACTAATAATGTAGCCGAGTGCTCTTTCCAGGAAGGTCGGGTGAGCAG
TGCAGAGGGCTGCCCACAGATTTTGCCCAGCAGTGACATCCTGGTACCGG
CGGGGATCGTTCGGCCAATCACATTGCGAGCCCGAAACTTGCCCCAGCCT
CAGTCTGGACAGAAGAACTATGAGTGCGTCTTTAACATCCAGGGAAAAGT
GCAGCGTATTCCTGCGGTCCGCTTCAACAGTTCCTGCATCCAGTGTCAGA
ACACCTCGTACTGGTATGAAGGGAACGAGATGGGGGATCTGCCTGTGGAT
TTCTCCATCGTGTGGGACGGTGACTTTCCCATCGACAAACCCTCATCCATG
AGAGCTCTCCTGTATAAGTGTGAGGCTCAGAGGGACAGCTGTGGACTATG
TCTGAAGGCTGACAGCACATTTGAGTGTGGCTGGTGTTTGGCCGATAAGA
AGTGTCTCCTAAAGCAACACTGTCCATCAGCCGAACACAACTGGATGCATC
AGGGACGACGCAACATTCGCTGCAGCCATCCGCGCATTACCAAGATTCGT
CCTCTGACGGGCCCGAAAGAAGGAGGCACACGCGTCACCATTGAAGGGG
```

AGAATCTGGGGCTGCAGGTTCGAGAAATCACTCACGTGCGTGTGGCTGGA
GTTCGCTGTAACCCTGCTGCAGCTGAATACATCAGCGCTGAGAGGATTGT
GTGTGATATGGAGGAGTCCCTGATGTCCAGTCCTCCCGGAGGTCCGGTG
GAGCTGTGTATCGGAGACTGCAGCGCTGAGTACAGGACTCAATCCACACA
GACTTACTCCTTTGTGATGCCGAGCTTCAGTCGAGTGCGCCCTGAGAAAG
GCCCGGTGTCCGGCGGGACGAGGCTGACCATCTCAGGCCGACATCTGGA
CGCCGGCAGCGCTGTGACCGTGTTTTTGGCCCAGGAGGAGTGTCTGTTC
GTCAGGAGGACGGTGCGTGAGATTGTGTGTGTGACGCCTCCATCAGCTTC
AGGATCTGGACCTTCATCTGTGAAGCTGTTTATTGATAAAGCAGAGATCAC
CAGCGACACCCGCTACATCTACACTGAAGACCCAAATATCTCCACCATCGA
GCCCAACTGGAGCATCATCAACGGCAGCACAAGCCTCACGGTCACAGGAA
CCAACCTGCTCACCATTCAGGAGCCCAAAGTCAGAGCCAAATATGGAGGA
GTGGAGACCACAAACATCTGTAGTCTGGTCAATGACTCTGTGATGACGTG
CTTGGCTCCGGGCATCATCTACACTAAACGTGAGGCTCCAGAAAGCGGCG
TTCACCCGGACGAGTTCGGCTTCATCCTGGATCACGTCTCTGCCCTCCTC
ATCCTCAACGGGACTCCGTTCACTTACTATCCCAACCCGACCTTTGAACCT
CTTGGGAATGCCGGGATTCTGGAGGTCAAACCAGGATCACCCATCATCCT
GAAGGGCAAGAACCTGATTCCTCCTGCGCCTGGGAACATCCGTCTGAATT
ACAGCGTGACGATCGGAGAAACGCCCTGCCTGCTAACAGTCTCTGAATCT
CAGCTGCTCTGCGATTCGCCAGATCTGACCGGAGAACAGCGAGTGATGAT
TCTTGTCGGCGGTCTGGAATATTCCCCCGGAATGCTTCACATTTATTCGGA
CAGCACTCTCACGTTGCCTGCCATCATCGGGATCGGAGCAGGTGGAGGA
GTCCTCCTCATCGCCATCATCGCTGTGCTCATCGCTTACAAGCGCAAGAC
GCGGGACGCCGACCGCACACTCAAACGCCTGCAGCTGCAGATGGACAAC
CTGGAGTCCCGGGTTGCGCTGGAGTGCAAGGAAGCATTCGCTGAGCTGC
AGACAGACATCCAAGAGCTGACGAATGACATGGACGGTGTGAAAATCCCT
TTCCTGGAGTATCGTACCTACACCATGAGAGTGATGTTCCCTGGCATCGAG
GAGCACCCGGTTCTGAAGGAGCTGGACTCTCCAGCTAATGTGGAGAAGGC
CCTGCGCTTGTTCAGTCAGCTGCTGCACAACAAGATGTTCCTGCTGACCTT
CATCCACACGCTGGAGGCGCAAAGGTCCTTCTCCATGCGGGATCGTGGCA
ATGTGGCCTCCCTCCTCATGGCGGCACTGCAGGGACGGATGGAGTACGC
CACTGTGGTTCTCAAACAGCTGCTAGCCGACCTGATCGAGAAGAACTTGG
AGAACCGAAACCACCCTAAACTACTGCTTAGACGAACTGAATCTGTGGCAG

AGAAGATGCTCACCAACTGGTTCACGTTCCTTCTGCACCGCTTCCTCAAGG
AGTGTGCGGGCGAGCCTCTGTTTATGCTGTACTGTGCTATAAAACAGCAG
ATGGAGAAAGGCCCCATAGACGCCATCACAGGAGAGGCCAGATACTCCCT
GAGCGAAGACAAGCTCATCCGACAGCAAATCGACTACAAGCAGCTGACGC
TGATGTGTATTCCTCCTGAAGGAGAAGCCGGGACAGAAATCCCTGTTAAG
GTGCTAAACTGTGACACGATCACTCAGGTGAAGGACAAGCTGTTGGACGC
TGTTTATAAAGGCATCCCGTACTCGCAGAGACCACAGGCGGACGACATGG
ACCTGGAATGGCGGCAGGGTCGACTGACCAGAATCATCCTCCAAGATGAA
GACGTCACCACAAAGATCGAGAGCGACTGGAAGAGACTGAACACACTGGC
ACATTACCAGGTGACAGATGGGTCTTTGGTGGCTTTGGTTCAGAAGCAAGT
ATCCGCTTACAACATCGCCAACTCTTTCACGTTCACTCGCTCTCTCAGTCG
ATACGAGAGCCTCTTGAGGACGTCCAGTAGTCCAGACAGCCTGCGCTCCA
GGGCTCCCATGATCACTCCTGACCAGGAAACGGGTACCAAACTCTGGCAC
CTGGTGAAGAACCATGAGCATGCAGACCAGCGGGAAGGAGACCGCGGCA
GCAAGATGGTGTCTGAGATTTACCTCACACGCTTACTAGCTACCAAGGGCA
CTCTGCAGAAGTTTGTGGACGATCTGTTTGAGACGGTCTTCAGTACAGCTC
ACCGCGGCAGCGCTCTCCCGCTGGCCATCAAATACATGTTTGATTTCCTG
GATGAACAGGCGGACAAGAGGCAGATCACCGACCCAGACGTACGGCACA
CCTGGAAGAGCAACTGCCTTCCTCTGCGGTTTTGGGTCAACGTGATCAAA
AACCCTCAGTTTGTGTTTGACATCCACAAGAACAGTATTACAGATGCCTGT
CTGTCGGTGGTGGCTCAGACATTTATGGACTCCTGCTCCACGTCTGAGCA
TCGTCTGGGAAAAGACTCTCCGTCAAACAAACTGCTCTACGCTAAAGACAT
CCCCAACTACAAGAGCTGGGTGGAGAGATATTACCGTGACATCAGCAAGA
TGCCAAGTATCAGTGATCAGGATATGGATGCCTATCTGGTCGAGCAGTCTC
GTCTCCATGGCAACGAGTTCAACACACTGAGCGCGCTCAGTGAACTGTAT
TTCTACATCAACAAGTACAAAGAAGAGATTTTGACAGCGCTGGACAGAGAC
GGTTACTGTCGCAAACACAAGCTACGACACAAACTGGAACAAGCCATTAAC
CTGATGTCTGGCAGCAGCGGATCCTA-3'

7.3.3 Restriction enzyme map for *plexinA3* overexpression construct

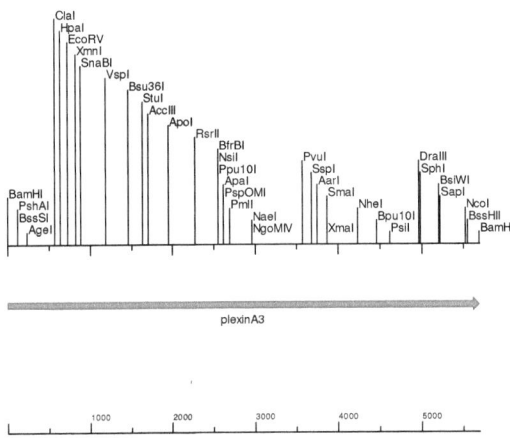

plexinA3 with BamHI sites (5692 bps)

7.4 Publications

Reimer M.M., Kuscha V., Wyatt, C., Sörensen I., Frank R.E., Knüwer M., Becker T.*, Becker C.G.* (2009) Sonic hedgehog is a polarized signal for motor neuron regeneration in adult zebrafish, J Neurosci. 29 (48):15073-82.

Reimer M.M., Sörensen I., Kuscha V., Frank R.E., Liu C., Becker C.G., Becker T. (2008) Motor neuron regeneration in adult zebrafish. J Neurosci. 28 (34):8510-6.

Feldner J., **Reimer M.M.**, Schweitzer J., Wendik B., Meyer D., Becker T., Becker C.G. (2007) PlexinA3 restricts spinal exit points and branching of trunk motor nerves in embryonic zebrafish. J Neurosci. 27 (18):4978-83.

* These authors contributed equally

Die VDM Verlagsservicegesellschaft sucht für wissenschaftliche Verlage abgeschlossene und herausragende

Dissertationen, Habilitationen, Diplomarbeiten, Master Theses, Magisterarbeiten usw.

für die kostenlose Publikation als Fachbuch.

Sie verfügen über eine Arbeit, die hohen inhaltlichen und formalen Ansprüchen genügt, und haben Interesse an einer honorarvergüteten Publikation?

Dann senden Sie bitte erste Informationen über sich und Ihre Arbeit per Email an *info@vdm-vsg.de*.

Sie erhalten kurzfristig unser Feedback!

VDM Verlagsservicegesellschaft mbH
Dudweiler Landstr. 99   Telefon +49 681 3720 174
D - 66123 Saarbrücken   Fax   +49 681 3720 1749
www.vdm-vsg.de

Die VDM Verlagsservicegesellschaft mbH vertritt

Printed by Books on Demand GmbH, Norderstedt / Germany